# Practical Physiology & Pathology

(With Solved Oral Questions for Medical Students)

*By*

**DR. MAHESH SHARAN VERMA**

D.H.M.S. (HONS.) PAT, F.P.T. (MUZ.)

Head of the Deptt. of Physiology
B.N.M. Homoeo Medical College & Hospital
S A H A R S A
(Bihar)

**B. JAIN PUBLISHERS (P) Ltd.**
NEW DELHI—110055

**PRACTICAL PHYSIOLOGY AND PATHOLOGY**

Reprint Edition: 2006

No part of this book may be reproduced, stored in a retrieval system or transmitted, in any form or by any means, mechanical, photocopying, recording or otherwise, without any prior written permission of the publisher.

© Copyright with the publisher.

**Price: Rs. 15.00**

*Published by Kuldeep Jain for*
## B. Jain Publishers (P) Ltd.
1921, Street No. 10, Chuna Mandi,
Paharganj, New Delhi 110 055 (INDIA)
*Phones:* 91-11-2358 0800, 2358 1100, 2358 1300, 2358 3100
*Fax:* 91-11-2358 0471; *Email:* bjain@vsnl.com
*Website:* www.bjainbooks.com

*Printed in India by*
## J.J. Offset Printers
522, FIE, Patpar Ganj, Delhi - 110 092
*Phones:* 91-11-2216 9633, 2215 6128

ISBN: 81-7021-057-7
BOOKCODE: BV-2538

# FOREWORD

I have gone through this book and found that this is an important work of Dr. M.S. Verma, D.H.M.S. (Hons.), Head of the Department of Physiology of our institution. The book is most beneficial for the students of D.H.M.S., B.A.M.S., G.U.M.S., and other medical science and medical practitioners.

I wish him more success in future for such type of work.

MOIZUDDIN

B.A, D.H.M.S. (Hons.) Pat., R.F.P.T. (Muz.)
P.Gr. P.M.R. (Bombay)

Ex-Senior House Surgeon, R.B.T.S.H.
Medical College & Hospital (Muz.)

Vice-Principal, B.N.M. Homoeo-Medical
College & Hospital, Saharsa.

Active Member, International
Homoeopathic Medical League, Geneva.

Vice-President, H.M.A.I., Bihar.

Secretary, All India Pharmaceutical
Committee.

# PREFACE

Now this small book is in your hand with great pleasure Homoeopathic students and practitioners will be greatly benefited as it is specially compiled for them. But undoubtedly it has got equal importance among all other systems of medicine. The book is designed with a purpose to solve all problems of examinees which arise specifically at the moment of practical and oral examination in the subject of Physiology and Pathology.

The book contains a large number of laboratory investigations and also a number of solved critical questions. Therefore, the book as a whole is divided in three parts :

Part First : Physiology and Pathology Practical.

Part Second : Physiology Solved Oral Questions.

Part Third : Pathology Solved Oral Questions.

The questions are collected on each and every chapters of different books of Physiology and Pathology.

The language is very simple and all subjects are described in a clear manner so as to make easy to understand for every student.

I expect that this book will perform a best duty to fulfil the demand of students appearing in the examination of Diploma or Degree courses in medical science.

**M.S. Verma**

# CONTENTS

## PART I

**Physiology and Pathology Practical**

### Chapter 1

| | *Pages* |
|---|---|
| **The Examination of Urine** | **1—15** |
| Physical examination | 3 |
| Chemical examination | 6 |
| For chlorides | 7 |
| For protein (Albumin) | 7 |
| For blood | 9 |
| For sugar | 10 |
| For bile pigment | 12 |
| For Urobilinogen and Urobilin | 13 |
| For bile salt | 13 |
| For acitone bodies | 14 |
| Microscopical examination of urine | 14 |

### Chapter 2

| | |
|---|---|
| **Laboratory Examination of Stool** | **16—20** |
| Naked eye examination | 17 |
| Chemical examination | 18 |
| Reaction test of stool | 18 |
| For occult blood | 19 |
| For bile pigment | 20 |
| Microscopical examination of stool | 20 |

( vi )

**Chapter 3**

| | Pages |
|---|---|
| **Laboratory Examination of Blood** | **21—42** |
| Estimation of percentage between plasma and blood cells | 21 |
| Haemoglobin estimation | 22 |
| Total R.B.C. count | 25 |
| Total W.B.C. count | 27 |
| Differential count of W.B.C. | 29 |
| Total count of platelets | 32 |
| Determination of E.S.R. | 33 |
| Examination of blood for parasites | 35 |
| Method of recognition of parasites | 36 |
| Method of sternal puncture | 39 |
| Examination of blood for L.D. bodies or diagnosis for Kala-azar | 40 |
| Examination of blood for V.D.R.L. | 41 |

**Chapter 4**

| | |
|---|---|
| **Laboratory Examination of Sputum** | **43—46** |
| Naked eye examination of sputum | 43 |
| Microscopical examination | 45 |
| For abnormal constituents | 46 |

**Chopter 5**

| | |
|---|---|
| **Examination of Seminal Fluid** | **47—50** |
| Method of collection of semen | 47 |
| Physical examination of semen | 48 |
| Microscopical examination of semen | 48 |
| Total count of spermatozoa | 49 |

|  | Pages |
|---|---|
| Examination of motility of sperm | 49 |
| Morphological examination of sperm | 50 |

### Chapter 6

**Examination of Cerebrospinal Fluid (C.S.F.)**    51—54

| | |
|---|---|
| Physical examination of C.S.F. | 51 |
| Chemical examination of C.S.F. | 52 |
| Microscopical examination of C.S.F. | 53 |
| Examination of parasites and bacteria | 54 |

## PART II

**Physiology Oral Questions and Answers**    55—84

| | |
|---|---|
| Cell and tissue | 57 |
| Blood | 59 |
| Circulatory system | 62 |
| Respiratory system | 64 |
| Digestive system | 66 |
| Metabolism | 69 |
| Vitamins | 69 |
| Excretory system | 73 |
| Endocrinology | 74 |
| Reproductive system | 77 |
| Nervous system | 82 |

## PART III

**Pathology Oral Questions and Answers**    85—132

### Chapter 1

    **Terminology**    87

( viii )

|  | Pages |
|---|---|
| **Chapter 2** | |
| **Necrosis and Gangrene** | **93—101** |
| Infections | 94 |
| Inflammation | 96 |
| Immunity | 98 |
| Circulatory disturbances | 101 |
| Disorders of blood | 105 |
| Tumour or Neoplasm | 108 |
| Oedema | 111 |
| Allergy | 112 |
| Exudation and suppuration | 113 |
| Fever or Pyrexia | 115 |
| Regeneration | 117 |
| Degeneration | 118 |
| Infiltration | 120 |
| Atrophy | 121 |
| Hypertrophy | 122 |
| Reparative process | 123 |
| **Chapter 3** | |
| Helminthology | 127 |
| **Chapter 4** | |
| Protozoology | 129 |
| **Appendix** | 133 |

# PART I

# Physiology and Pathology Practical

CHAPTER 1

# THE EXAMINATION OF URINE

The examination of urine is conducted by the following methods :
- (*i*) Physical examination.
- (*ii*) Chemical examination.
- (*iii*) Microscopical examination.

## (I) PHYSICAL EXAMINATION

It is done by naked eye, hence it is also known as naked eye examination. In the physical examination of urine the following points are observed in the laboratory :

- (*a*) Colour and transparency.
- (*b*) Odour.
- (*c*) Specific gravity (Sp. gr.).
- (*d*) Total quantity.
- (*e*) Reaction.
- (*f*) Naked eye characters of deposits.

(*a*) **Colour and transparency.** It is seen by naked eye. The normal colour of urine is yellowish (pale).

**Note.** The variation in the colour depends upon the following factors :

| Factors | Colour |
|---|---|
| (1) Presence of blood in small quantity | Smoky |
| (2) Presence of blood in large quantity | Brownish or red |
| (3) Presence of bacteria | Cloudy |
| (4) Presence of calcium phosphate | Surface layer is milky or white |
| (5) Presence of bile | Brown or dark brown |
| (6) In jaundice | Deep yellow. |

Normally the urine is transparent when freshly passed.

(b) **Odour**. The normal odour of urine is "Arometic" but after passing some times, it changes into "Ammoniacal."

(c) **Specific gravity (Sp. gr.)**. "Urinometer" is an instrument by which the Sp. gr. of the urine is measured.

The normal Sp. gr.—1010 to 1020 but some suggests 1010 to 1025.

*Identification of urinometer.* It is made up of glass and graduated from downward to upward from 0 to 60 or 1000 to 1060. Here 0 (zero) means 1000.

**Note**. The calibration of the urinometer is checked by seeing that it reads 1000 or 0 when placed in distilled or pure water.

*Method of Measuring the Sp. gr. of urine.* The urine is taken in a clean and dry test-tube, 2/3rd. The test-tube should be wide as the urinometer may be placed in freely. Now the urinometer is placed in the test-tube urine and the reading should be read.

The Sp. gr. of the urine varies in the following conditions :

| Conditions | Sp. gr. |
|---|---|
| (1) By cooling the room temperature | Increases |
| (2) In diabetes melitus | ,, |
| (3) In presence of protein | ,, |
| (4) In diabetes incipidus | Falls. |

**Note.** One percent protein increases 3 point Sp. gr. of urine.

(d) **Total quantity.** The normal quantity of urine in 24 hours varies widely from 700 to 2500 ml.

The average quantity is 1500 ml or about 50 ozs. (ounces).

**Note.** 1. Increased quantity of urine is known as "Polyuria".

2. Decreased quantity of urine is known as "Oliguria".

3. Loss (absent) of urine is known as "Anuria".

The quantity of urine increased in 24 hours in the following conditions :

| Conditions | Quantity |
|---|---|
| (i) Day time | More |
| (ii) At night | Less |
| (iii) In renal failure | |
| In day time | Less |
| At night | More |
| (iv) Increased consumption of food or drink | More |
| (v) After exposure of cold | ,, |
| (vi) In both diabetes | ,, |
| (vii) High blood pressure | ,, |
| (viii) Hysteria | ,, |

| Conditions | Quantity |
|---|---|
| (*ix*) Profuse perspiration | Less |
| (*x*) Less fluid and more solid in diet | ,, |
| (*xi*) In the condition of fever | ,, |
| (*xii*) Vomiting | ,, |
| (*xiii*) Diarrhoea | ,, |
| (*xiv*) Concussion of the brain | ,, |
| (*xv*) Nephritis | ,, |
| (*xvi*) Low blood pressure | ,, |

(*e*) **Reaction**. Reaction of the urine is tested with the help of litmus paper.

In acid the blue litmus paper turns its colour in red and in alkaline, red litmus paper turns its colour in blue.

The normal reaction of the urine is slightly "Acidic".

(*f*) **Naked eye character of deposits**. The normal urine is perfectly clear and transparent fluid when examined just after micturition, but after some time it shows the deposits of mucous.

## (II) CHEMICAL EXAMINATION

The chemical examination of urine is applied for testing the following points :

(1) Examination of urine for chlorides.
(2) ,, ,, ,, ,, proteins (Albumin).
(3) ,, ,, ,, ,, blood.
(4) ,, ,, ,, ,, sugar.
(5) ,, ,, ,, ,, bile pigments.
(6) ,, ,, ,, ,, urobilinogen and urobilin.
(7) ,, ,, ,, ,, bile salt.
(8) ,, ,, ,, ,, acetone bodies.

## EXPERIMENT NO. 1

### **Examination of Urine for Chlorides**

*Requirements.*  20% Potassium chromate solution.
2.9% Silver nitrate solution.

Test-tubes, pipette, sample of urine.

*Procedure.* Ten drops of urine is taken in a small test tube with the help of pipette. The pipette is rinsed and one drop of the 20% Potassium chromate solution is added. Again the pipette is rinsed and 2.9% Silver nitrate solution is added drop by drop. Colour turns into brown from yellow. The test-tube is shaken properly after each addition.

*Observation.* The number of drops of the Silver nitrate solution used to change in the colour gives the concentration of chlorides in the urine.

Each drop indicates the gram of Sodium chloride presence per litre of urine.

Normally 5 to 15 gram Sodium chlorides are passed in 24 hours with urine.

## EXPERIMENT NO. 2

### **Examination of Urine for Protein (Albumin)**

This test is done in two ways :

(*a*) Sulpho-salicylic acid test.
(*b*) Boiling test.

### **Sulpho-salicylic acid test**

*Advantage.* This test is very reliable and does not require heat.

*Requirements.* Small test-tube, dropper, filter paper, 20% Sulpho-salicylic acid.

*Procedure.* The sample of urine is filtered by filter paper before proceeding to apply test for protein, it is necessary that the sample of urine should be absolutely clear, if not, please clear the urine by following methods in different circumstances :

(i) The urine is filtered with filter paper.

(ii) After filtering more than one, the turbid colour of the urine indicates the bacteria are probably present and they can be removed by long centrifugation or by shaking up the urine with powdered Barium carbonate and filtering.

(iii) Heat is applied to disappear the turbidity if the urates are the cause.

Now 5 ml. of filtered urine is taken in a clean dry test-tube and 6 drops of Sulpho-salicylic acid is added with the help of a clean dropper.

*Observation.* The formation of a cloud indicates the presence of Protein (Albumin).

## Boiling Test for Protein (Albumin)

*Requirements.* A test-tube, blue litmus paper, 10% Acetic acid, test-tube holder, spirit lamp.

*Procedure.* A small clean test-tube is filled upto two-third of urine. The reaction of the urine is tested with the help of litmus paper. If the urine is alkaline, 10% Acetic acid is added drop by drop till the urine becomes acidic. The test-tube is hold with the help of test-tube holder at the bottom and inclined at an angle. Then heat is applied by spirit lamp on the top of the test-tube one inch over the flame.

A cloudiness indicates the presence of either Protein or Phosphate. Now 10% Acetic acid drop by drop is added and heat is again applied.

*Observation.* (i) If the cloud disappears, indicates Phosphate.

(ii) If it persists, indicates the presence of Protein in the urine.

(iii) If the precipitate partially persists and partially disappears, it indicates the presence of both in small quantity.

If appearance of cloudiness is faint, it indicates the Albumin is in traces and it is reported as $1+$, $2+$, $3+$, $4+$, according to the opacity of the cloudiness.

## EXPERIMENT NO. 3

### Examination of Urine for Blood

It is done in the following ways :

(a) Guaiac test for the blood pigment.

(b) Occultest tablets.

(c) Caustic soda test.

(a) **Guaiac Test**

*Requirements.* Sample of urine, test-tube, Tincture of guaiacum, Ozonic ether.

*Procedure.* Some urine is taken in a test-tube and two drops of Tincture of guaiacum is added into it. A partial white precipitate appears. Now same quantity (quantity equal to taken urine) of Ozonic ether is added without shaking.

*Observation.* Appearance of a blue colour at the line of junction of the fluids, indicates the presence of blood pigment in the urine.

## (b) Occultest Tablets Test

*Requirements.* Test paper, sample of urine, occultest tablets, pure water.

*Procedure.* One drop of urine is placed in the centre of the test paper and one occultest tablet is placed on moist area of the paper. Two drops of pure water is droped on the tablet.

*Observation.* Appearance of blue colour within two minutes on the paper around the tablet, indicates the presence of blood in the sample of urine.

## (c) Caustic Soda Test

*Requirements.* Test-tube, sample of urine, Caustic soda, spirit lamp, test-tube holder.

*Procedure.* Some urine is taken in the clean and dry test-tube and some Caustic soda is added to make sufficient alkaline. Now heat is applied by sprit lamp with holding the test-tube by test-tube holder.

*Observation.* A brownish red sediment and above the fluid formation of green colour shows the presence of blood in the urine.

**Note.** (i) Presence of blood in the urine in large quantity shows the colour of urine red and this condition is known as "Haematuria".

(ii) If blood pigment present is known as "Haemoglobinurea".

## EXPERIMENT NO. 4

### Examination of Urine for Sugar

This experiment is done in three ways but first and second

methods are commonly used in the laboratory but the first method is most convenient and sensitive.

(1) Benedict's test.
(2) Clinitest.
(3) Fehling's test.

### (1) Benedict's Test for Sugar in Urine

(This test is commonly used in practice).

*Requirements.* Benedict's reagent (qualitative), sample of urine, test-tube, test-tube holder, dropper, spirit lamp.

*Procedure.* 5 c.c. of Benedict's reagent is taken in a clean and dry test-tube and 8 drops of urine is added into it by the help of a clean dropper.

Now heat is applied for 2 minutes by spirit lamp holding the test-tube by test-tube holder. Now it is allowed to cool.

*Observation.* The changes in the colour of the Benedict's solution is observed as follows :—

(a) If the colour remains unchanged on cooling, indicates sugar is absent.
(b) If the colour turns greenish with yellow precipitates at the bottom of the test-tube the sugar is in trace (1+)
(c) If it turns in yellow, the sugar is more (2+).
(d) If turns in red, the sugar is more (3+).
(e) If turns in brick' red uniformally (4+) means enough (profuse) sugar is present in the urine.

### (2) Clinitest Method for Sugar in Urine

*Requirements.* Sample of urine, test-tube, dropper, distilled water, clinitest tablet, spirit lamp.

*Procedure.* 5 drops of urine is taken in a test-tube with a clean dropper. Again 10 drops of clean water is added in the urine and one clinitest tablet is dropped into it. Then heat is applied for 15 seconds. Boiling occurs and changes in the colour of the solution is observed.

*Observation.* (i) If the colour is not changed, it indicates the absence of sugar.

(ii) If changed in green, $\frac{1}{2}\%$ sugar is present.

(iii) If colour becomes orange, $2\%$ sugar is present.

## EXPERIMENT NO. 5

### Examination of Urine for Bile Pigment

This experiment is done in three ways.

(1) A clear filter paper is taken and sample of urine is filtered from it. A few drops of Nitric acid is put on the moist area.

*Observations.* Appearance of green colour indicates the presence of bile pigment.

(2) This test is also done by Ntiric acid hence it is known as "Nitric Acid Method".

3 c.c. Nitric acid is taken in a test-tube and urine passed gently from the side of the test-tube.

A ring of green colour at the junction of Nitric acid and urine indicates the presence of bile pigment in the urine.

(3) About 10 c.c. of urine is taken in a test-tube and 3-4 drops of Loffler's Methylene blue or Iodin is dropped into the test-tube.

Change of colour from blue to green indicates the presence of bile pigment in the urine.

## EXPERIMENT NO. 6
### Examination of Urine for Urobilinogen and Urobilin

*Requirements.* Sample of urine, Ehrlich's Aldehyde reagent.

*Method.* 5 c.c. urine is taken in a test-tube and 1 c.c. of Ehrlich's Aldehyde reagent is added to it. After 3 to 5 mintues changes in the colour is observed.

*Observation.* If the colour changes in distinct red colour, it indicates that the urobilinogen is present in the sample of urine.

**Note.** Bilirubin is the secretion of the liver into the bile. In the intestine, it is reduced by the intestinal bacteria and converts into urobilinogen and urobilin. Maximum portion of the urobilinogen and urobilin reabsorbed and circulates with blood and very small quantity is excreted through the urine, normally. It increases in the urine in following conditions.

In hapatic dysfunction such as—Infective hepatitis and diffuse diseases of the liver. Absence of urobilinogen and urbilin indicates obstructive diseases of the liver. Such as—obstructive jaundice in which the obstruction is complete and so that no bile pigment is reaching in the intestine.

## EXPERIMENT NO. 7

### Examination of Urine for Bile Salt

*Requirements.* Sample of urine, test-tube, sulphur powder.

*Method.* Some quantity of urine is taken in a clean test-tube and Sulphur powder is sprinkled on the surface of the urine.

*Observation.* (a) If the sulphur powder remain on the surface, indicates the absence of the bile salt.

(b) If powder settles to the bottom passing through the urine, indicates the presence of bile salt.

## EXPERIMENT NO. 8
### Examination of Urine for Acetone Bodies

*Name of the test.* "Rothera's Test".

*Requirments.* Sample of urine, crystals of Ammonium sulphate, solution of Sodium nitroprusside, strong Ammonia solution and test-tube.

*Method.* 10 ml. of urine is taken in a test-tube and a saturated solution is made by adding the crystals of Ammonium sulphate in excess quantity. Then 2 c.c. of strong Ammonia solution is added.

Formation of a deep permagnate colour indicates the presence of Acetone and Aceto-acetic acid both.

Now Aceto-acetic acid is tested by "Gerhardt's Test" as follows :

The same urine is taken in a test-tube and 10% Ferric chloride solution is added dropwise. A precipitate forms, again disappears after adding the Ferric chloride in large quantity. The solution becomes brownish red which indicates the presence of Aceto-acetic acid.

### III Microscopical Examination of Urine

*General procedure.* The urine is taken in a clean test-tube and it is centrifuged by centrifusing machine. To facilitate the process an another test-tube is filled with same quantity of distilled water, it is put in the opposite hook of the centrifuging machine. If there are already two specimen to test, there is no need to take distilled water.

Now it should be kept in the machine for five to six minutes to contrifuge.

At the bottom of the test-tube some deposits are seen after centrifusing. These deposits are taken in very small quantity on a clean and dry plane glace slide and is covered with cover slip. Then it is put on the stage of microscope to examine the following :—

(1) Crystals.
(2) Pus cells.
(3) Red blood cells.
(4) Casts etc.

CHAPTER 2

# LABORATORY EXAMINATION OF STOOL

The laboratory examination of stool is done in three ways :
(1) Naked eye examination (Inspection).
(2) Chemical examination.
(3) Microscopical examination.

Before going through the laboratory it is our first and formost duty to know about the normal composition of stool (Faeces).

The normal stool is composed by the following :

(a) Water 65%.

(b) Solid 35%.

Indigestible Protein, particles of meat, cells and fibres of the vegetables etc.

(c) Bile pigment, enzymes, mucus and some other product of digestive tract.

(d) A number of bacteria.

(e) Indole, gases, fatty acid and skatole are found in the stool as a product of decomposition.

(*f*) Cells of the wall of the intestine.

(*g*) Some digested foods which are carried out before absorption.

## Naked Eye Examination of Stool

The following points should be considered in this heading. It is also named by "Physical Examination". Amount, colour, odour, form and consistancy, frequency etc.

*Amount.* Average normal quantity of stool in 24 hours is about 200 gms. but its variation depends upon the quality of diet.

Such as the quantity increases after taking the vegetarian meal in the diet.

*Colour.* The colour varies somewhat with the character of the diet.

Paler stool on milk diet and black stool after taking iron containing substances. Dark green due to spinach, green due to large doses of colomel.

## Pathological Alteration in Colour

(a) The colour of the stool is changed due to presence of accult blood.
(b) White and clay due to absence of bile.
(c) Greasy due to large amount of fat, pancreatic diseases and Tubercular peritonitis.
(d) Bright red and brown due to bleeding and some times persistant diarrhoea.
(e) Tar-like black due to perforation in the intestine.
(*f*) Green colour in children diarrhoea.

*Odour.* Peculiar offensive due to the presence of the product of decomposition. In vegetable diet it is less offensive.

Stool is very offensive due to absence of bile as in the case of jaundice.

## Form and Consistance

Normally semi-solid. In constipation it becomes hard and dry and in the case of diarrhoea it becomes watery.

## Frequency

Normally in 24 hours once or twice but it depends upon the quality of diet.

## Chemical Examination of the Stool

It is done for the following :

(1) Reaction.

(2) Occult blood.

(3) Bile pigment.

## EXPERIMENT NO. 1

### Reaction Test of Stool

For the purpose to know the reaction of stool the following requirements are required in the laboratory :

*Requirements.* Sample of stool, distilled water, blue and red litmus paper.

*Method.* The stool is made liquid by the addition of distilled water. Then blue and red litmus papers are put in the liquid of the stool.

*Observation.* (*i*) If the blue litmus paper turns in the red colour, indicates the reaction of the stool is acidic.

(*ii*) If the red litmus paper turns in blue colour, indicates the reaction is alkaline.

(*iii*) If there is no change in the colour of any litmus paper, indicates the stool is nutral in reaction.

**Note.** The strong acidic reaction indicates the fermentation and strong alkaline reaction indicates the putrefaction.

## EXPERIMENT NO. 2

### Examination of Stool for Occult Blood

It is done in two ways :

### (1) Haematest Tablet Test

*Requirements.* Sample of stool, test paper, Haematest tablet (Orthotolidin+Strontium peroxide), distilled water.

*Method.* A thin smear is made of the sample of stool and this smear is taken in the centre of the test paper. Then one tablet of Maematotest is placed on it and two drops distilled water dropped on the tablet.

Appearance of a blue colour in the smear at the side of the tablet within two minutes, indicates the blood is present in the stool.

### (2) Second Way

2 c.c. distilled water is taken in a test-tube and a piece of stool is dropped into it. Now it is pluged with cotton and heat is applied till boiling. Then same quantity of distilled water is added in it for the purpose of cooling. Afterwards 15 drops of Benzidine solution (some Benzidine +3 c.c. of Glacial acetic acid) in an another test-tube is taken and 5—6 drops of the prepared solution of the stool is added in it and shaked properly. Then 3 c.c. of 3% Hydrogen peroxide is added and shaked.

*Observation.* Appearance of a blue green or deep blue colour shows the presence of blood in the sample of stool.

## EXPERIMENT NO. 3

### Examination of Stool for Bile Pigment

*Requirements.* Corrosive sublimate conc. test-tube.

*Method.* Some stool is mixed with conc. solution of corrosive sublimate and kept in for 24 hours.

*Observation.* (1) If there is no change in colour, indicates that the bile pigment is not present.

(2) If the colour of the stool is changed into red colour, indicates the presence of bile pigment in the sample of stool.

### Microscopic Examination of Stool

The solid stool is liquified by mixing some distilled water for the purpose of microscopic examination.

With the help of one end of a match-stick one drop of this liquified stool is taken on the slide. Then it is covered with a cover slip and put under the microscope for examination.

Under this examination we have to look for the following :

(a) Muscle fibres.

(b) Mucous.

(c) Blood cells.

(d) Bacteria, worms and parasites.

(e) Yeast and other fungi.

(f) Starch granules.

(g) Vegetable particles.

(h) Connective tissue and elastic fibres.

(i) Fat and its derivatives.

(j) Oxalates, phosphates and Cholesterine crystals.

CHAPTER 3

# LABORATORY EXAMINATION OF BLOOD

The examination of blood is done under the following headings :

(1) Estimation of percentage between plasma and blood cells.
(2) Estimation of percentage of Heamoglobin.
(3) Total count of R.B.C.
(4) Total count of W.B.C.
(5) Differential count of W.B.C.
(6) Total count of platelet.
(7) Erythrocyte sedimentation rate.
(8) Examination of blood for parasites.
(9) Examination of blood for L.D. bodies.
(10) Examination of blood for V.D.R.L.

## EXPERIMENT NO. 1

### Estimation of Percentage Between Plasma and Blood Cells

*Introduction.* Normally the plasma is found in the blood about 52 to 55% and blood cells are 45-48%.

It is estimated by an instrument named as *Haematocrit*.

The heamatocrit is made up of glass capillary tube. It is graduated from downward to upward (10 to 100%) which shows the percentage. Bore of this tube is uniform and very narrow like capillary tube.

*Method.* The sample of blood treated with an anticoagulant is taken in the Haematocrit up to 100 marks. The blood is taken from the subject's vein with the help of a sterillized syringe. Now it is centrifuged with the help of a centrifuging machine at a speed of 3000 revolution per minute for 30 minutes. Then it is allowed in rest till the sediment of corpuscles are completely settled at the bottom of the Haemotocrit.

Now there is appearance of a margin at the middle of the Haematocrit in the blood as the blood has been divided in two layers. The lower layer, from the margin, indicates the blood cells and the upper layer from the margin, indicates the plasma of the blood which is read by graduation of the tube in percentage.

## EXPERIMENT NO. 2

### Haemoglobin Estimation

This experiment is done by two methods :

(1) By Halden's Maemoglobinometer.

(2) By Sahli's Haemoglobinometer.

Haemoglobin in thealthy body is 100% or 14·8 gm.

## Estimation of Haemoglobin (Hb) By Haldan's Haemoglobinometer

### Method of Taking the Specimen of Blood

Tip of any finger specially middle finger, or lobe of the ears or toes and in the case of infants heel is used for the purpose. At first the selected part is cleaned with antiseptic lotion. (Such as Calendula Q, Dettol, Tincture iodine, Hydrogen peroxide etc.). A needle (Pricking needle) is sterilised by any one of these antiseptic agents. Now the cleaned part is punctured with the cleaned needle. The first drop of blood is wiped with cotton gauge. The next drops are used for the experiment.

*Method.* 0·4% solution of Ammonia is taken in the diluting tube up to the 20 mark. The tip of the middle finger or any part of the body as indicated above is punctured with a sterilised needle and the capillary pipatte is filled up to the 20 mark very quickly (in between 2 to 5 minutes). The capillary pipatte should be clean and dry. Now this taken blood is gently blown into the diluting tube. After that the solution is carefully saturated with Carbon monoxide (CO) by the help of a capillary pipette or lumber puncture needle. Distilled water is now added dropwise with the help of a dropper and the colour is compared from the prescribed comparison glass tube containing Carboxy-haemoglobin. Water is added in this manner until the colour in the diluting and prescribed standard tubes appears identical. The tube should be matched in day light but not direct in sun light against a white paper at an angle of 45°. At last after matching the level of the dilution is read in the tube which indicates the percentage of Hb in blood.

In this method there is less chance of error. However, 5% error is found by skilled observer.

## Estimation of Haemoglobin by Sahli's Haemoglobinometer

*Introduction.* Sahli's Haemoglobinometer (Haemometer) is an instrument to measure the percentage of haemoglobin (Hb). It has two colouring tubes one each side. Its one side is marked in percentage from downward to upward and the other side is marked in grms. in the same manner. In the middle there is also a graduated tube having mark 20 which is named by diluting tube.

*Method.* The sample of blood is taken (from the middle finger of the left hand by the help of pricking needle after proper sterilisation of the finger and needle by antiseptic lotion) in Hb pipatte upto the marking in diluting tube, up to the mark 20, N/10 Hcl is taken. The blood of the pipatte is drawn into the N/10 Hcl of the diluting tube, the pipatte being rinsed out several times with the acid solution. Now the diluting tube is allowed to stand for five minutes. Then distilled water is added dropwise in the solution and change colour of the bloody solution is carefully matched from the side tubes. The addition of the distilled water is continued dropwise till the colour of solution matches from the colour of matching tubes.

Now reading is taken in percentage or grams.

Normal—100% or 14·8 grms.

## EXPERIMENT NO. 3

### Total R.B.C. Count

*Requirements.* Sample of blood, R.B.C. diluting fluid, R.B.C. pipatte, cover slip, glass slides, clinical microscope, counting chamber.

(1) *R.B.C. diluting fluid.* It is a solution of :

    (a) Sodium citrate      1 gm.
    (b) Formalin      1 gm.
    (c) Sodium chloride      0·6 gm
    (d) Distilled water      100 ml.

*Or*

(2) Hayem's Solution :

    (a) Sodium chloride      1 gm.
    (b) Corrosive sublimate      0·5 gm
    (c) Sodium sulphate      5 gm.
    (d) Distilled water      200 ml.

*R.B.C. Pipatte.* It is a graduated glass tube which one side is fixed with a rubber tube and other side is free and pointed somewhat. Just above the middle of the tube, dilated portion is known as bulb. The pipatte has got three graduations. One the stem and below the bulb there are marking of 0·5 and 1 and just above the bulb there is a third mark of 101.

*Counting chamber.* It is a special type of glass slide prepared for counting of R.B.C. and W.B.C. Its ruling area consists of a square millimeter, the central square millimeter is divided into 25 groups and each group separated by triple lines. Each area of each group is again divided into 16 small squares and the length of each side of small square is 1/20th mm.

*Cover slip.* It is very thin slide of glass only for covering the counting chamber at the time of counting.

## Method of R.B.C. count

Sample of blood is taken from the finger (see page 23) in the R.B.C. pipatte up to the mark 1 and then in the same pipatte R.B.C. diluting fluid is taken up to the mark 101. Thus the ratio between blood and fluid is 1 : 100, but the dilution of the blood will be 100, because the last part of fluid remains locked up in the stem is not available for dilution. If blood is taken up to mark 0·5 the ratio will be 1 : 200. Now one drop of diluted blood from the R.B.C. pipatte is dropped in the central area of counting chamber and covered by cover slip.

The height of the cover slip from the surface of the counting chamber is 1/10th mm. So the volume of each small square is $1/20 \times 1/20 \times 1/10 = 1/4000$th cu. mm. (1/20 mm. is the length of one side of one small square).

Red cells are counted in the 5 groups of 16 small squares.

So, $5 \times 16 = 80$ small squares.

So, the volume of 80 small square
$$= 1/4000 \times 80 = 1/50 \text{ cu. mm.}$$

To avoid twice counting of a cell (corpuscles) those on a line are counted only, as on the top and left lines or on the bottom and right lines.

The number of red cells per cu. mm.

$$= \frac{\text{Number of cells counted} \times \text{dilution} \times 4000}{\text{Total number of small squares counted}}$$

( 27 )

Suppose, we find 900 R.B.C. in one small square so,

$$\frac{900 \times 100 \times 4000}{80}$$

$= 900 \times 100 \times 50 = 4500000$

$= 4\cdot 5$ million per cu. mm. of blood R.B.C.

For the purpose of convenience and to avoid such big equations, we count the R.B.C. of all 80 small squares and multiply them the number with 10,000. This gives the number of total R.B.C. per cu. mm. of blood.

## EXPERIMENT NO. 4

### Total W.B.C. Count

*Requirements.* Sample of blood, W.B.C. pipatte, counting chamber, W.B.C. diluting fluid cover slip, clinical microscope.

*W.B.C. pipatte.* The W.B.C. pipatte has a white beed in its bulb where as R.B C. pipatte has red beed. In W.B.C. pipatte above the bulb there is a mark of 11 whereas in R.B.C. pipatte 101. Other construction is similar in both except the bulb of W.B.C. is smaller than the bulb of R.B.C.

*W.B.C. diluting fluid.* Commonly is made as follows :

Glacial acetic acid—1·5 ml.

1% solution of Gentian violet in water—1 ml.

Distilled water—98 ml. and a small quantity of thymol (Thymol is used to avoid the formation of moulds).

### Method of Counting

The sample of blood is taken (see page 23) in the W.B.C. pipatte up to mark 1 and in the same pipatte, up to mark 11, W.B.C. diluting fluid is taken. Thus the ratio between the

( 28 )

blood and diluting fluid is 1 : 10. Now it is mixed thoroughly and then a few drops dropped off as there is only the diluting fluid below the bulb.

Now one drop of this solution is taken in the counting chamber and covered by cover slip and put under the microscope for counting.

Normally the total count of W.B.C. is 5000 to 8000 per cum. mm. of blood. The white cells are counted in the four big squares and in central ruled area on both sides of the counting chamber. The cells touching the boundary lines are not counted. Thus the number is multiplied by 10.

| | |
|---|---|
| One side of the big square | $= 1$ mm. |
| One side of the small square | $= 1/4$ mm. |
| So, the area of 1 small square | $= 1/4 \times 1/4$ $= 1/16$ square mm. |
| Height of the cover slip is | $= 1/10$ mm. |
| So, the volume of one small square | $= 1/16 \times 1/10$ $= 1/160$ cu. mm. |

and volume of 16 small squares or one big square
$$= 1/160 \times 16 = 1/10 \text{ cu. mm.}$$

*For example.* Let in one big square or 16 small squares we count the W.B.C. as 60. It means

$$1/10 = 60$$
or $$1 = 60 \times 10$$
or $$1 \text{ cu. mm.} = 600$$

as the dilution is 1 : 10

Hence in 1 cu. mm. of blood total count of W.B.C.
$$= 600 \times 10 = 6000.$$

Meaning 6000 W.B.C. per cu. mm. of blood.

For practical purpose we adopt the following formula :

$$\frac{\text{Number of cells counted} \times \text{dilution} \times 10}{\text{Number of big squares counted}} = \text{total count}$$

$$= \frac{60 \times 10 \times 10}{1} = 6000 \text{ per cu. mm. of blood.}$$

## EXPERIMENT NO. 5
### Differential Count of W.B.C.

Normal differential count of W.B.C. are :

(1) Neutrophil   3000—6000 per cu. mm. of blood or 60—70%
(2) Eosinophil   150—400      ,,           ,,      or  1—4%
(3) Basophil     0—100        ,,           ,,      or  0—1%
(4) Lymphocytes 1,500—2,700   ,,           ,,      or 25—30%
(5) Monocytes    350—800      ,,           ,,      or  5—10%

## How to Collect the Specimen of Blood from Vein

For the purpose to make film or for other experiment the sample of blood is taken from the vein of the subject. Over the middle of the biceps muscle of the forearm a tourniquet is applied to stop the flow of vein but not of artery. At the bend of the elbow, the skin is painted with 2% Sodin in spirit or cleaned with ether or spirit. By the help of left hand of operator the skin is rendered tense. Almost parallel with the paient's arm a needle attached syringe is hold in the right hand and the needle with the bevel upwards is inserted into a prominent vein of the patient's arm. The required quantity of blood is then drawn up into the siringe and the torniquet is removed, from the arm, before the needle is withdrawn, as otherwise a haematoma may form. In some cases as soon as the needle is inserted the tourniquet is removed so the free flowing blood is taken. As soon as the needle is withdrawn a swab is placed on the site of puncturer and the patient is advised to hold his forearm firmly flexed against his arm for two minutes or more.

In some cases a vein in the forearm or wrist may prove more convenient than one at the elbow.

Now the taken blood should immediately be placed in a suitable container. Different anticoagulants are needed for particular purposes when uncoagulated blood is required. Such as—Heparin, Sequestrene (EDTA) are generally and commonly used for most haematological investigations.

**Method to Make Films**

A clean and dry glass slide is put on a level surface holding it with thumb and index finger of the left hand. One drop of sample of blood (see page 23) is dropped or placed on one end of slide. The narrow end of a second slide is placed in the drop and held there till the blood has spread across it. This second slide is placed on the first slide inclining at an angle of 45°. Then it is drawn slowly over the whole length of first slide without applying pressure in between two surfaces.

The thinner layer of spread blood is the resulting film. Now it is dried by being waved rapidly in the air to prevent undue shrinkage of the cells.

**Method to Stain a Film**

The film is stainded by any one of the following methods :

(1) Janner's stain.

(2) Leishman's stain.

*Janner's stain.* After drying the film (see above) a few drops of Janner's stain solution are poured on and the film is covered by watch glass to prevent precipitation and evaporation of the stain. Now it is poured off within one to four minutes and rinsed in distilled water till pink (10 to 15 seconds). Then it is dried by waving in the air. Now it is mounted in xylol balsam.

In successful film the red cells are brownish red, bacterial filarial and malarial parasites—blue ; nuclei are blue ; platelets, the granules of polynuclear cells and myelocytes—red.

Basophils—dark violet.

The Janner's stain is composed by 0.5% solution of a specially prepared crystalline compound of methylene blue and eosin in pure methyl alkohal.

*Leishman's stain.* This is very simple method introduced firstly by Romanowsky. The dried film is well covered with Leishman's stain. After one minute, distilled water in double quantity is carefully added and mixed with the stain. After seven minutes, it is poured off and the film is covered by distilled water for two or three minutes. Now the water is washed off with fresh distilled water and then it is dried by waving in air.

## Method to Examine a Film

A drop of xylol balsam or liquid paraffin is placed on the film and oil immersion objective of the microscope is used after low power (Low power objective is used for the general examination of film). It is observed that whether the film is properly spread and whether the R.B.C., W.B.C. and platelets are present in the normal proportion. Then counting is made.

## Method of Differential Count of W.B.C.

A well prepared film is taken and counting is made by high power or oil immersion lens until 200 have been counted.

For normal differential count (see page 29).

## EXPERIMENT NO. 6

### Total Count of Platelets

It is done by any one of the following two ways:
 (i) Indirect Method.
 (ii) Direct Method.

### Indirect Method

The lobe of the ear is cleaned with ether or spirit and a big drop of Ethylenediamine tetra-acetic acid or 3% Sodium citrate in normal saline is placed on the cleaned surface. A pin puncture is made through the drop and the blood is taken on a clean slide to prepare the film. (Ethylenediamine tetra-acetic acid or 3% Sodium citrate in normal saline acts as an anticoagulant through which blood is taken).

Now the film is covered by cover glass, ringed with petroleum jelly and put under the moving stage of microscope having squared eye piece and 1/12 inch oil immersion lens. The number of platelets and red cells are counted in several fields and determination is made by the ratio of platelets and red cells. It is counted as the number of platelets present per 100 red cells in each field.

Normal total Number of platelets—250,000 to 453,000 per cu. mm. of blood.

The number falls below 150,000, produce-Thrombocytopenia.

The number increases above 450,000 produces Thrombocythaemia.

*Direct Method.* In this method measured amount of anticoagulant alongwith dye are taken in a clean pipatte and these are mixed thoroughly. Its counting is made by the help of counting chamber along with R.B.C. and determination is made by the ratio of platelets and R.B.C.

## EXPERIMENT NO. 7

### Determination of Erythrocyte Sedimentation Rate

*Introduction.* When blood is mixed with an anticoagulant and allowed to stand in a tube, the corpuscles are found to sediment gradually at the bottom of the tube, this phenomenon is called sedimentation and the rate of sedimentation is known as "Erythrocyte Sedimentation Rate".

In healthy person or in health, the red cells of the blood settle down very little on standing and the rate of sedimentation is very low but in certain diseases the red cells agglutinate into large clump and the rate of sedimentation becomes very high. Such as in the case of tuberculosis, cancer, chronic cough, asthma, chronic fever etc., the sedimentation rate increases.

In some cases the sedimentation rate is reduced also. Such as, polycythaemia and congestive heart failure. Physiological variations are as follows :

It is lowest in newborn, 0.0-2.0 mm. per hour.

In children slight higher 9.0 mm. per hour.

In pregnancy and in old age E.S.R. increases.

*The sedimentation rate* is estimated by following two ways.
(1) Westergren Method.
(2) Wintrobe Method.

*Westergren Method.* In a small syringe 0.4 ml. of 3.8 percent Sodium citrate solution (Anticoagulant) is taken and by the help of same syringe 1.6 ml. veinous blood is taken and it is mixed thoroughly. Now this blood is sucked into a Westerngren tube upto 0 (zero) mark. The tube is put in a vertical position and its lower end is placed on a rubber cap and the upper end reamins open. The height of the tube is graduated in mm. It is left free for one hour and the reading is noted. Again after one hour the reading is noted and the sedimentation rate is clarified by the following formula :

$$\frac{\text{Reading of 1st hour} + \text{Reading of 2nd hour}}{2}$$

Suppose, 1st hour reading is 15 mm.
2nd hour reading is 30 mm.

So, E.S.R. $= \frac{15+30}{2}$ ; (2 = Time)

or $\frac{45}{2}$ or 22·5 mm. per hour.

The Westergren method is most commonly used in laboratory. By this method the normal E.S.R. is 3 to 7 mm.

*Wintrobe Method.* In this method the oxaleted blood is taken in the tube of haematocrit upto the mark 100 and allowed to stand vertically for one hour or two hours. Other procedure is adopted as Westergren. The advantage of this method is that the same sample of blood may be allowed in the same tube for the determination of percentage of plasma and blood cells.

Increased sedimentation rate indicates the persistant infection in the body but it is not important for the purpose of diagnosis. During the treatment if the E.S.R. begins to fall, indicates the improvement in the condition and the prognosis is certainly good.

## EXPERIMENT NO. 8

### Examination of Blood for Parasites

The following important parasites generally may be found in the blood.

Such as, the parasites of malaria, trypanosomes, spirochaetes of relapsing fever, microfilariae of several types, leishman-donovan bodies etc.

For the purpose to look these parasites, the preparation of the fresh film is required in which some parasites *e.g.* trypanosomes and microfilariae may be seen alive and moving. The slide (film) may be prepared thin or thick or both according to need.

Procedure of thin slide preparation has already been described.

*Method of preparation of thick films.* The finger or lobe of the ear is pricked with a needle adopting full aseptic precautions and the first drop of blood is wipped away. The next drop of blood is taken on the centre of a clean, dry and greasless slide. Now the drop of the blood is spread on the centre with the help of a triangular needle until print can be clearly seen through it while it is still wet. Then the film is allowed to dry thoroughly for two hours or if possible overnight by a precaution to protect it from dust. This dried film is now placed into a beaker containing distilled water to desolve out the haemoglobin by a gentle movement (but the slide, if is handled roughly the film may become detached). In about five minutes, when the haemoglobin desolves or the film is colourless and opaque, it is removed from the water and allowed to dry. Thereafter it is stained with Jenn's stains or Leishman's stains, in the same way as the thin films.

Thick film is inspected rapidly under the 2/3 or 1/6 or if necessary 1/12th, and then examined systematically under oil immersion lense.

Following tables show the nature of infection and need of films (thick and thin) respectively.

| Nature of Infection | Need of Thick or Thin Film |
|---|---|
| (1) Malarial infection | Both films. |
| (2) Trypanosomiasis | Thick or thin. |
| (3) Leishman donovan | Thick or thin but it may fail. Blood culture methods are frequently sucessful. |
| (4) Filarial infection | Only thick film. |

## Method of Recognition of Parasites

### (1) **Malarial Parasites** :

| Types | Condition Produces |
|---|---|
| (a) Plasmodium falciparum | Malignant tartian malaria |
| (b) Plasmodium vivax | Benign tartian malaria |

(c) Plasmodium malaria          Quartan malaria

(d) Plasmodium ovale            Combination of plasmodium vivax and malariae.

In some cases mixed infection may be present.

For the diagnostic purpose of malaria, thick film should be used for the detection of parasites and thin film for their identification.

The blood is taken when the temperature of the patient is raised.

### (2) Infection of Trypanosomes (Trypanosomiasis):

The examination of the blood is of less importance. In this case the gland is punctured by a thin sterile needle and the fluid is examined.

The moderate size of hypodermic needle is inserted in the gland and the small amount of gland fluid passes into the needle, there is no need of suction. Then the needle is withdrawn and its contents (whatever a small quantity) placed on a clean glass slide. Now it is examined unstained and the thin film is stained with Laishman's stain and examined under oil immersion. Fresh film are examined under the 2/3 and then under the 1/6 objective.

In advanced cases the films are made from the deposit of centrifused cerebro-spinal fluid (C.S.F).

### (3) Diagnosis of Filarial Infection

The adult filarial worms or macrofileriae are found in the lymphatic system or connective tissues. The larvae or microfilariae are present in the blood stream.

Following are the types of filaria parasites :

| Types | Site of Presence | Disease Produces |
|---|---|---|
| (a) Filaria Bancrofti- | Found in blood stream only at night | Filariasis with irregular fever, lymphangitis and various forms of elephantiasis. |
| (b) Filaria loa-loa | Found in blood-stream only in day time. | Loaisis, characterised by transient red painful swellings named by calabar swelling |
| (c) Filaria perstans | Found in blood stream equally by day and night | No recognised pathogenic effects. |

In the suspect cases of filaria, the specimen of blood is taken at 8 a.m.. noon, 4 p.m., and again at 8 p.m. midnight and 4 a.m.

Fresh unstained film is used for the detection of filariae.

The thick film is used if the larvae are scanty and thin stained in that case if they are plentiful for their identification.

Leishman's stain can not stain the microfilariae therefore it is stained as follows :

After dehaemoglobinizing and drying the thick film or thin film is mixed with Methyle alcohol for 5 minutes and rinsed in water. Now Ehrlich's mixture (Haematoxyle) is used in which the film is placed and heat is applied until steam rises and an occasional buble is seen. The heat is removed for a few minutes and again heat is applied. This process is repeated several times in a period of 15 minutes. During this period if slide becomes dry the stain is added certainly. When slide becomes cool, it is immersed in water to wash off the stain. After that it is placed in running tap water until a blue colour appears. Now the slide is dried and then mounted in canada balsam and then it is examined under microscope.

### (4) Leishman—Donovan bodies infection :

It may be looked for in blood or in that material which are are obtained by sternal, gland, spleen or liver puncture.

The sternal puncture is the simplest and safest method, but in some cases the parasites may be found by re-examination of stained blood film.

The blood culture method is also frequently successful when direct imcroscopical examination fails.

### Method of Sternal Puncture :

*Indication.* It is a method applied to examine the marrow in anaemia and leukaemia to find out the certain parasites.

Such as, Leishmaniasis, carcinometosis, myelomatosis and Gaucher's diseases, etc.

For this purpose a special instrument is supplied named as Troctar and cannula, fitted with an adjustable stop. It is as made as a hypodermic syringe may fits in the top of cannula.

*Method.* The local surfaces (in and out) of the sternum are anaesthetized by 2% novocain or gesicain without adrenaline. Now, opposite the 2nd or 3rd intercostal space over the middle of the sternum, a small incision is made in the skin with a tenotome. After that the sternum is punctured by trochar with a boring motion by applying some force. Now the trochar is withdrawn and the hypodermic syringe is inserted and a few drops as far as possible, red fluid is aspirated. Now the canula is removed and a swab is placed over the incision then syringe is disconnected.

The film is made immediately as far as possible on slide from the fluid of the cannula.

The film may be stained with Jenner's or Leishman's stain.

## EXPERIMENT NO. 9

### Examination of Blood for L.D. Bodies or Diagnosis for Kala-Azar (Leishmaniases)

*Requirements.* Syringe of 5 c.c., formalin, blood serum.

This test is called Aldehyde test.

*Method.* 5 c.c. blood is taken from the vein with the help of a syringe. It is allowed to coagulate in a clean test-tube.

2 c.c. serum is taken with the help of a dropper or a syringe from the coagulated blood in a next test-tube. Now in the serum one drop of formalin is added and it is thoroughly mixed. Then this mixture is allowed to stand for 24 hours.

*Observation.* If the serum becomes thick, opaque and milky indicates the test is positive.

(1) If these changes occur in serum within an hour, it means test is strongly positive ($4+$).

(2) If changes take place with 2 hours, it indicates $3+$.

(3) If changes take place within 3 hours, it indicates $2+$.

(4) If changes take place within 4 hours, it indicates $1+$ (weakly positive).

**Note.** Positiveness of the test shows the infection of Leishman—Donovan (L.D.) bodies. (Leishmaniases) partial solidification may occur in leprocy, syphilis, tuberculosis, but spacity never occurs in these conditions.

## EXPERIMENT NO. 10

### Examination of Blood for V.D.R.L.

### (Venerial Disease Research Laboratory)

**Tube Method :**

*Requirements.* (a) V.D.R.L. antigen. (b) Buffered saline. (c) Unbuffered saline (d) Blood serum. (e) Khahn shaker

*Introduction.* This test is done for serodiagnosis of syphilis.

(a) V.D.R.D. antigen is a mixture of the following chemical reagents :

| | |
|---|---|
| (i) Lecithin | 0.25 per cent |
| (ii) Cholesterol | 0.90 per cent |
| (iii) Cardiolepin | 0.04 per cent. |

(b) Buffered saline is a mixture of the following chemicals :

| | |
|---|---|
| (i) Formaldehyde | 0.5 ml. |
| (ii) Sodium chloride | 10.0 ml. |
| (iii) $KH_2PO_4$ | 0.18 gr. |
| (iv) $Na_2HPO_4, H_2O$ | 0.1 gr. |
| (v) Equa dest | 1.0 litre. |

(c) Unbuffered saline—One per cent NaCl.

*Method.* 0.4 ml. buffered saline is taken dropwise in a clean test-tube in which 0.5 ml. V.D.R.L. antigen is added dropwise frequently within few seconds and during this period bottle is continuously rotated. After this both chemicals are thoroughly mixed by rotating the bottle vigorously. Now 3.6 ml. of one per cent Sodium chloride is added and mixed and allowed to stand for 5 minutes.

This preparation can be used within 2 hours only. Now 0.5 ml. of serum is taken in a tube and 0.5 ml. diluted antigen is mixed in it. The tube is shaken for 5 to 6 minutes in a kahn shaker and then it is spun at the rate of 2000 rotation per minute for 10 minutes.

After this the observation is made immediately.

*Observation.* (1) Appearance or a clear of faintly turbid medium in the form of aggregation indicates positive.

(2) Appearance of a turbid swirl on gentle shaking—Weakly positive or negative.

(3) Appearance of turbid on border line only—Negative..

CHAPTER 4

# LABORATORY EXAMINATION OF THE SPUTUM

The materials which are coughed out from the respiratory organs, except the nosal mucus and secretion of the oral cavity, are named as sputum. Morning sputum is the best sample for examination purpose.

The examination of sputum is done under the following headings :

(1) Naked eye examination or physical examination.

(2) Microscopical examination.

**Naked Eye Examination**

In this examination we have to consider the following points :

($i$) Quantity ($ii$) Colour
($iii$) Consistancy and appearance ($iv$) Odour.

**Quantity**

The perfect quantity of expectorated sputum is not so essential to know but in the case of pulmonary tuberculosis, pulmonary oedema, lung abscess, bronchiecatasis, pneumonia, bronchitis etc., the quantity increases.

However, in normal case it is expectorated few ml. in morning.

## Colour

The normal sputum is transparent and colourless.

The significance of colour is as follows :

| Colour | Case |
|---|---|
| (a) Red or reddish. | Blood or blood pigment |
| (b) Rusty | Pneumonia |
| (c) Prune juice | Drunkard's pneumonia |
| (d) Grey or black | Coal-dust worker. |
| (e) Blood stained | Bronchiectasis |
| (f) Bright red | Pulmonary tuberculosis or trauma to be chest. |
| (g) Yellowish green | Advanced tuberculosis and chronic bronchitis. |

## Consistancy and Appearance

The sputum of a subject may be tenacious, serous, purulent or mucopurulent but it differs in different conditions.

In the case of gangrene, bronchiectesis and pulmonary abscess, if the sputum is kept in the test-tube for several hours, it may separate into two or three layers and often it is seen that the top layer is frothy and mucoid, the second is watery and opaque and the bottom layer contains pus, tissue and bacteria.

## Odour

Normally it is ineffective but in some pathological conditions it becomes offensive.

Such as, in the case of bronchiectasis, gangrene and abscess of the lung and in the case of malignant diseases of the lungs usually it becames offensive.

## Microscopical Examination

The microscopical examination of the sputum is done for the investigation of bacteria and non-bacterial abnormal constituents.

## Investigation of Bacteria

The sample of sputum is taken on a clean slide and film is made. The film is stained by two methods :

(a) Gram method.

(b) Ziehl-Neelsen's method.

## Gram method :

This method is used for the clarification and identification of gram-positive and gram-negative micro-organism.

The film of the sputum is made on a slide and fixed. Some crystal violet is placed on it then Cram iodin is placed on it after two minutes and washed with plane water. Decolourization of the slide is made with the help of spirit or ether. Again it is stained as counter strain with safranin. Again it is washed and dried. Then placed under microscope for examination.

The bacteria retaining violet colour are gram-positive.

The bacteria retaining safranin colour are gram-negative.

For example, of gram-positive and gram-negative bacteria.

Gram-positive-streptococci, staptylococci, pnumococci bacillus diphtheriae, mycobaeterium tuberculosis, mycobacterium leprae.

## Gram-negative

Haemophilus influenzae and haemophilus pertussis.

## Ziehl-Neelsen's method

The method is applied for the identification of acid-fast bacilli such as mycobacterium leprae, mycobacterium smergmatis and mycobacterium tuberculosis.

The sputum is taken uniformally on a slide and heat is applied from below putting Ziehl-Neelsen's stain on the slide. Heat is applied till smoking. When smoking starts heat is removed for 4 to 6 minutes to dry. After drying it is washed with water. The decolourization is made with 25% Sulphuric acid and again washed. This process is repeated again and again till the slide is clear. Then it is counter stained with malachil green or methylene blue and again washed and dried. Then it is placed under microscope for examination.

Acid fast bacilli takes previous Ziehl-Neelsen's stain (red colour) and most of the other bacteria takes methylene blue.

## Examination of Sputum for Abnormal Constituents

The examination is done without staining the film for the identification of the following constituents :

(a) Malignant cells—Found in pulmonary tumour.

(b) Elastic fibres—Present in the case of tuberculosis of lung abscess, etc.

(c) Charcot leyden crystals—In the sputum of asthma patients.

(d) Curschman's spirals—Bronchial asthma.

(e) Pigmented cells—These are fine colourless crystals with sharp ends.

(f) Fibrinous casts—Found in pneumonia.

CHAPTER 5

# EXAMINATION OF SEMINAL FLUID

For the purpose to investigate the causative factor of sterility this examination is done in the laboratory. The cause of male sterility is mainly due to either decreased quality or decreased quantity of the spermatozoa in the semen.

Examination is done under three headings :

(a) Physical Examination.

(b) Microscopical Examination.

(c) Morphological Examination.

## Method of Collection of Semen

The following points should be considered :

(1) Before one week the subject is advised to avoid sextual intercourse (copulation).

(2) The semen is collected in a clean, dry and wide-mouth test-tube by masturbation.

(3) The hour of emission (discharge) should be noted.

(4) For collecting the semen, condom is not suitable because the deleterious effect of rubber may disturb the spermatozoa.

### (a) Physical Examination

Under this heading the following points are considered :
- (i) Quantity or volume.
- (ii) Colour and odour.
- (iii) Viscosity.
- (iv) Reaction.

**Quantity.** Normal about 4 ml.

It is measured in a small graduated cylinder. The quantity or volume may reduced after first emission if next is discharged at short interval from the same donor.

### Colour and odour

Normally it is whitish with a typical seminal smell.

Foul smell indicates the suppurative infection. Yellow colour also indicates the suppuration. Red colour indicates the haemorrhage in the vesicles.

### Viscosity

If it is freshly ejaculated—highly viscous. After 10 to 30 minutes of ejaculation—it liquefies.

It has a property of self-liquefaction. The absence of liquefaction may inhibit or disturb the motility of spermatozoa which may interfere the fertilization and it may appear as a cause of sterility.

### Reaction

Slightly alkaline.

### (b) Microscopical Examination

Microscopical examination is done for the following :
1. Total count of spermatozoa.
2. Motility of spermatozoa.

# TOTAL COUNT OF SPERMATOZOA

**Requirements :**

W.B.C. pipette, counting chamber, cover glass, seminal diluent.

**Introduction :**

For W.B.C. pipette, counting chamber and cover glass (see Blood examination).

*Seminal diluent.* It is fluid composed by the following chemicals :

(i) Distilled water          100·0 ml.
(ii) Formalin          1·0 ml.
(iii) Sodium carbonate          5·0 ml.

*Method.* Semen is taken in W.B.C. pipette upto mark 0·5 and then diluent upto mark 11. The other procedure and method of counting and calculation of the number of spermatozoa are done as in the case of W.B.C. and the resulting number is multiplied by 1000. It is indicated in per cubic centimetre.

*Normally.* Total number of spermatozoa is 150 to 200 million per cu. centimeter.

## EXAMINATION OF MOTILITY

One drop of semen is placed on a well cleaned slide. A ring is made by vaseline around the drop of the semen and covered with a clean cover slip by applying some pressure on cover slip. Now it becomes airtight. Then it is put under microscope and motile spermatozoa are counted from high power lens. It is reported in percentage. Forward moving spermatozoa are counted specially and considered motile.

After discharge of semen it is done at once at the interval of 3, 6 and 24 hours.

### Morphological Examination

In this examination stained semear is made by the sample which is diluted for the sperm count. Diluted semen may be stored for several days to examine the morphology of spermatozoa.

The method of preparation of stained smear of semen is same as blood but it is stained with Wright's stain or Giemsa's stain or by the Gram method but some prefers the Haematoxylin and eosin staining also. Then it is placed under microscope and oil immersion lense is used.

In the morphology the following points are considered :

(i) Double head.

(ii) Double tail.

(iii) Premature sperms.

(iv) Macrocephalic.

(v) Microcephalic.

Pus cells, epithelial cells and gram-negative cocci are also reported.

CHAPTER 6

# EXAMINATION OF CEREBROSPINAL FLUID (C.S.F.)

The examination of cerebrospinal fluid is done by the following ways :

(a) Physical Examination :
    (I) Pressure.
    (II) Colour and consistancy.

(b) Chemical Examination for Protein.

(c) Microscopical Examination :
    (I) Cell count.
    (II) Parasites and Bacteria.

### (a) Physical Examination

(I) *Pressure.* The pateint is advised to lie in left lateral position. A water manometer attached lumber puncture needle is taken and lumber puncture is made by the needle. Now the C.S.F. allowed to come out dropwise and pressure is noted by water manometer. *In the case of normal pressure C.S.F. runs out at the rate of 60 drops per minute.* It may run out rapidly under increased pressure as in the case of hydrocephalus and meningitis.

The normal pressure of C.S.F. is 60 to 120 mm. $H_2O$.

11 *Colour. and consistancy* : It is clear like water, colourless and with-out clot.

In tubercular meningitis the fluid is clear, colourless but has a spidery clot (cob-web) on standing.

In different conditions the colour and consistancy differs as follows :

| Conditions | Colour and Consistancy |
|---|---|
| (1) Coccal meningitis | Hazy, turbid or purulent and dense clot. |
| (2) Spinal blockage | Clear and golden yellow, dense clot. |
| (3) Old Haemorrhage | Blood stained or same in the case of spinal blockage. |
| (4) Jaundice | Yellow. |

But in the case of G.P.I., Tabesdorsalis, meningovascular syphilis, encephalitis, poliomyelitis, and cerebral tumour, there is no change in colour and consistancy.

### (b) Chemical Examination for Protein

This test is done by two methods : *1st Method*—One ml. C.S F. is taken in clean test-tube and it is shaken. There is appearance of froth of excess protein which does not disappear in five minutes.

Now five drops of 2% Salicyl-sulphonic acid is added resulting the formation of a white precipitate, while in the case of normal C.S.F. precipitate appears with stronger acid solution.

To know the quantity of Protein, 2.0 ml. C.S.F. is mixed with 0.3 ml. of 30% Trichlor acetic acid and heat is applied till boiling. After boiling it is left for one hour. It becomes opaque. The opacity is compared with a series of standard tubes ranging from 0˙02 to 0˙01%.

*2nd Method.* This test is named as Pandy's test for globulin (Protein).

1˙0 ml. of 10% Aqueous phenol (Pandy's reagent) is taken in a test-tube and few drops of C.S.F. are added.

The excess protein produces a cloud. There is no cloudy appearance in normal fluid but it remains clear.

In normal condition 0˙02% protein is found in C.S.F.

0˙30% in tubercular and coceal meningitis.

0˙30% to 4˙00% in spinal blockage, syphilis, encephalitis, cerebral tumour and poliomyelitis.

## MICROSCOPICAL EXAMINATION OF C.S.F.

(I) **Cell count**

10 ml. C.S.F. is mixed with 1 ml. of diluting fluid (Methyle violet+Acetic acid) and blood cells are counted by counting chamber as in the case of blood.

In normal condition.

Lymphocytes—2 or 3 per field.

Polymorph —Absent.

In acute infective meningitis the polymorphs are highly increased.

In acute and chronic infective cases—small lymphocytes present. The marked increased in lymphocytes (20 or 100 per field) is found in the case of tabesdorsalis and G.P.I. In tubercular meningitis, the polymorphonuclear is present.

## II Examination for Parasites and Bacteria

5·0 ml. C.S.F. is taken in a test-tube for centrifugation for 5 minutes. The supernatant fluid is decanted and films are made on clean glass slides from the deposits. Now film is dried in air and stained with the following stain—Gram's stain, Methylene blue, Leishman's stain and Ziehl—Neelsen's stain according to the requirement and the slide is placed under microscope.

# PART II

# Physiology Oral Question and Answer

# CELL AND TISSUE

**Q. 1.** Define Physiology.

**Ans.** Physiology is a branch of medical science which deals with the normal function of the different parts of the living body during rest and activity.

**Q. 2.** Who was the discoverer of Physiology?

**Ans.** It was discovered by a French physician named Jean Fernel in 1552.

**Q. 3.** What was the origin of physiology?

**Ans.** It was originated from a Greek word "Physiologikos". It means discourse on natural knowledge.

**Q. 4.** What is cell?

**Ans.** Cell is the structural and functional unit of the multicellular living body which is capable of maintaining all the processes of life independently.

**Q. 5.** What is protoplasm?

**Ans.** The living substance of animals and plants is termed as protoplasm.

**Q. 6.** Who was the discoverer of cell?

**Ans.** The cell was discovered by Robert Hook.

**Q. 7.** What are the types of cell division?

**Ans.** There are three types of cell division:
(*a*) Mitosis, (*b*) Meiosis, (*c*) Amitosis.

**Q. 8.** Define tissue.

**Ans.** It is the collection of cells which are similar in their structure and functions.

**Q. 9.** What are the types of tissue?

**Ans.** There are four types:
    (a) Epithelial tissue.   (b) Connective tissue.
    (c) Muscular tissue.   (d) Nervous tissue.

**Q. 10.** What are the types of epithelial tissue?

**Ans.** According to the shape and layers of cells, epithelial tissues are broadly divided into two groups:

(a) Simple epithelial tissue.

(b) Compound epithelial tissue.

**Q. 11.** What are the types of simple epithelial tissue?

**Ans.** There are five varieties of simple epithelial tissues. These tissues are found in one layer of cells and they are of five types:

(a) Pavement, (b) Cubical (c) Columner, (d) Ciliated (Columner and cubical) and (e) Glandular.

**Q. 12.** What are the types of compound epithelial tissue?

**Ans.** There are five types of compound epithelial tissues. They are found in more than one layer in the body:
(a) Transitional, (b) Stratified squamous cornified, (c) Stratified squamous non-cornified, (d) Stratified columnar, (e) Stratified columnar ciliated.

**Q. 13.** What are the types of connective tissue?

**Ans.** There are ten types of connective tissue:
    (a) Areolar tissue     (b) Adipose tissue
    (c) White fibrous tissue   (d) Yellow elastic tissue
    (e) Reticular tissue     (f) Blood and Haemopoitic tissue

(g) Cartilage tissue  
(h) Jelly-like tissue  
(i) Osseous tissue  
(j) Reticulo-endothelial tissue.

## BLOOD

Q. 14. Define blood.

Ans. It is red, thick, opaque and alkaline fluid in which liquid intercellular substance and blood cells are found. It comes under the gread of connective tissue.

The liquid intercellular substance is known as "Plasma".

Q. 15. What are the cells which are present in blood ?

Ans. There are three types of cells :
- (a) White blood corpuscles (W.B.C.) or Leucocytes.
- (b) Red blood corpuscles (R.B.C.) or Erythrocytes.
- (c) Platelates or Thrombocytes.

Q. 16. What are the total counts of blood cells ?

Ans. (a) Total count of W.B.C.—6000 to 8000 per c. mm. of blood.

(b) Total count of R.B.C. :     Per cmm of blood
- (i) In adult male 5 million       ,,    ,,
- (ii) In adult female 4.5 million  ,,    ,,
- (iii) In infants 6 to 7 million   ,,    ,,
- (iv) In foetus 7.8 million        ,,    ,,

(c) Platelets—250,000 to 450,000 per cu. mm. of blood.

Q. 17. What is the percentage of plasma and blood cells in the blood ? Name the instrument by which the percentage of plasma and cells are determined in the blood.

Ans. Plasma 55%, Cells 45%.

The instrument by which the percentage of plasma and cells is determined is known as "Haematocrit."

Q. 18. What are the proteins present in blood ?

Ans. There are four types of proteins which present in the blood :

(a) Serum albumin.

(b) Serum globulin.

(c) Fibrinogen.

(d) Prothrombin.

Q. 19. What are the functions of blood ?

Ans. (1) Blood acts as a transport of repiratory gases.

(2) It acts as a transport of nutrition.

(3) It maintains :

(a) Water balance.

(b) Acid-base balance.

(c) Ion balance.

(4) It regulates the body temperature.

(5) By forming antibody and having phagocytic, action, it acts as a defence in the body.

(6) It regulates blood pressure.

(7) It acts as a vehicle for vitamins and hormons.

(8) It guards against bleeding by the property of coagulation.

Q. 20. What do you mean by coagulation of blood?

Ans. The change in the state of blood from liquid to semi-solid jelly-like is known as coagulation.

Q. 21. What do you mean by blood volume?

Ans. It means the total amount of blood in circulation and in blood stores.

Q. 22. What is the normal blood volume in human body?

Ans. It is 1/11th part of the body weight or 2.5 to 4.0 litres per sq. metre of body surface.

Q. 23. What do you mean by haemoglobin?

Ans. It is the red pigment of blood having 96% globin (protein) and 4% heam (iron).

Q. 24. What is the normal quantity of haemoglobin in blood?

Ans. It 14.5 gm. or 100%.

Q. 25. What are the varieties of W.B.C.? Give their percentage and absolute number.

Ans. W.B.Cs. are broadly classified into two groups as follows:

(A) Agranulocytes:

   (i) Lymphocytes—25-30%, No. 1500-2700 per cu. mm. of blood.

   (ii) Monocytes—5-10%, No. 350-800 per cu. mm. of blood.

(B) Granulocyte:

   (i) Neutrophil—60-70%, No. 3000-6000 per cu. mm. of blood.

(ii) Eosinophil—1-4%, No. 150-350 per cu. mm. of blood.

(iii) Basophil—0-4%, No. 0-100 per cu. mm. of blood.

Q. 26. What do you mean by blood bank ?

Ans. It is a specialised medical centre which collects and stores human blood for transfusion.

## CIRCULATORY SYSTEM

Q. 27. What do you mean by circulatory system ?

Ans. It is a well-recognised transport system of the body by which the blood being circulated within a closed system under different pressure gradients, created by the pumping mechanism of heart which acts as a central pump.

Q. 28. What do you mean by arteries and veins ?

Ans. Arteries are those blood vessels which act as carrier of blood from the heart to the different parts of the body.

Veins carry the blood from the different parts of the body to the heart.

Q. 29. What do you mean by cardiac cycle ?

Ans. The cyclical repetition of the various changes in the heart in each and every beat is named by cardiac cycle.

The time required for one complete cardiac cycle is known as cardiac cycle time.

It is 0·8 second.

**Q. 30.** What do you mean by blood pressure ?

**Ans.** It is a lateral pressure applied by blood on the walls of the blood vessels while flowing through it at the time of systolic and diastolic action of the heart.

**Q. 31.** What are the types of blood pressure ?

**Ans.** There are four types of blood pressure :

(1) Systolic blood pressure (S.P.).
(2) Diastolic blood pressure (D.P.).
(3) Pulse pressure (P.P.).
(4) Mean pressure (M.P.).

*Systolic blood pressure* : During the systolic action of the heart (contraction), the pressure applied by blood, laterally, on the walls of blood vessels is named by systolic pressure.

The lateral pressure applied by blood on the walls of blood vessels, laterally, during diastolic action of the heart is known as diastolic pressure.

The difference between systolic and diastolic pressure is known as pulse pressure.

Such as—if systolic pressure—120 mm. Hg.
Diastolic pressure—80 mm. Hg.

So pulse pressure $= 120 - 80 = 40$ mm. Hg.

Mean pressure is the arithmetic mean of systolic and diastolic pressure.

Such as—
$$\frac{120+80}{2} = \frac{200}{2}$$
$$= 100 \text{ mm. Hg.}$$

In adult relation between the systolic, diastolic and pulse pressure is as follow :
SP/DP/PP = 120/80/40 = 3/2/1.

**Q. 32.** Name the instrument by which the blood pressure is measured.

**Ans.** Sphygmomanometer.

**Q. 33.** What do you mean by circulation time ?

**Ans.** It is the time required by one molecule of blood to flow from one point to another point in the circulation.

**Q. 34.** What do you mean by velocity of blood ?

**Ans.** It is a rate of blood flow through a given vessel in the body.

**Q. 35.** What do you mean by pulse ?

**Ans.** In the arterial walls expansion and elongation which passively produced by the pressure changes during the systolic and diastolic action of ventricles of the heart is known as pulse or radial pulse.

## RESPIRATORY SYSTEM

**Q. 36.** What are the advantages of nasal respiration ?

**Ans.** Advantages are as follow :

(1) Due to presence of nasal mucus the bacteria and dust particles caught up and thereby removed.

(2) The air becomes cool and obtains moisture from the nasal mucus.

(3) The odour of the inspired air can easily be taken due to presence of sense organ of smell in the nose and inspirable air is allowed to pass in.

Thus purified air passes down the respiratory organs.

**Q. 37.** Define the following terms :

Ans. (a) Tidal volume.
(b) Inspiratory reserved volume.
(c) Inspiratory capacity.
(d) Expiratory reserved volume.
(e) Residual volume.
(f) Functional residual capacity.
(g) Total lung volume or total lung capacity.
(h) Vital capacity.
(i) Dead space.

Ans. (a) *Tidal volume.* During quiet breathing the volume of air that can be taken in and given out is called tidal volume. It is about 500 ml.

(b) *Inspiratory reserved volume.* The volume of air that can be taken in by forced inspiration.
It is about 2000 ml. to 3300 ml.

(c) *Inspiratory capacity.* It is the volume of air that can be taken in during maximum inspiratory effort.
Nearly 3250 ml. to 3500 ml.

(d) *Expiratory reserved volume.* After normal expiration the volume of air that can be given out by forced expiration. It is about 1000 ml.

(e) *Residual volume.* The volume of air that can not be expelled out even by forced expiration without opening the chest wall.

It is about 1200 ml.

(f) *Functional residual capacity.* After normal expiration some volume of air persists in the lungs is called functional residual capacity.

(g) *Total lung volume.* It is volume of air persists in the lungs after a maximum inspiration.

It is about 5000 ml. to 6000 ml.

(h) *Vital capacity.* It is that volume of air that can be given out by forced expiration after taking a forced inspiration.

(i) *Dead space.* The amount of air which is locked up in the air passages.

## DIGESTIVE SYSTEM

Q. 38. Define digestion.

Ans. It is a biochemical process by which the complex food particles are broken down into simple unitary fragments for absorption and assimilation.

Q. 39. What are the types of food taken in diet ?

Ans. There are six types of foods which are taken in diet Such as : Protein, fat, carbohydrate, vitamins, salts and water.

Q. 40. What are the types of protein, carbohydrate and fat which are taken in diet ?

Ans. **Types of proteins :**

(a) Various types of albumins and globulins.

(b) Nucleoprotein.

(c) Caseinogen of milk.
(d) Collagen and gelatin.
(e) Mucin.
(f) Elastin.

## Types of carbohydrates

(a) *Polysaccharides.* Starch, glycogen, dextrin and cellulose.

(b) *Disaccharides.* Lactose, Maltose, Sucrose,

(c) *Monosaccharides.* Glucose and Fructose.

## Types of Fats

(a) Neutral fats.
(b) Phospholipids.
(c) Cholesterides.
(d) Free cholesterol.
(e) Fatty acids.
(f) Glycerol.

Q. 41. What are the types of digestive juices ?

Ans. There are five digestive juices :

(a) Saliva.
(b) Gastric juice.
(c) Intestinal juice.
(d) Pancreatic juice
(e) Bile.

Q. 42. Give the need of different types of digestive juices.

Ans. Digestion is carried out by enzymes and all types of enzymes are not present in any one digestive juice. So the presence of various types of digestive juices are essential for complete digestion of food.

Q. 43. What do you mean by enzymes?

Ans. Enzymes are soluble organic substances act as catalysts and manufactured by living cells but their functions do not depend upon the life of the cells.

Q. 44. Give the function of saliva?

Ans. (1) It keeps the mouth moist and helps in speech.

(2) It helps in the process of mastication and deglutition of the food-stuff.

(3) It cools the hot food.

(4) It dilutes the irritant substance.

(5) It does not allow to grow the bacteria inside the oral cavity.

(6) Ptyalin acts on boiled starch and converts into maltose.

(7) Maltase acts on maltose and converts into glucose.

(8) It helps in the excretion of urea, heavy metals, thiocyanates and certain drugs.

(9) It helps in the sensation of taste.

(10) It helps : Water balance and temperature balance.

(11) It acts as :

(a) Buffer.

(b) Bacteriolytic.

Q. 45. Give the functions of gastric juice.

Ans. There are five functions :

(a) Pepsin with the help of Hydrochloric acid (Hcl) digests protein and converts into peptone.

(b) Caseinogen of milk is coagulated by rennin which is present in the gastric juice.

(c) Fat is digested up to some degree by the action of gastric lipase.

(d) Hydrochloric acid acts as antiseptic and causes some dydrolysing action on food-stuffs.

(e) It excrets toxins, heavy metals and certain alkalies.

Q. 46. What do you mean by absorption ?

Ans. The process by which the end product of digestion enter the blood stream passing through the intestinal epithelium.

## METABOLISM

Q. 47. What do you mean by metabolism ?

Ans. It is a biochemical reaction which takes place in the body after absorption of food materials and before excertion.

It includes two processes :

(a) Anabolism (Synthesis).

(b) Catabolism (Breakdown).

## VITAMINS

Q. 48. Define vitamins.

Ans. These are organic compounds found in diet in different quantity to maintain the physiological process of the life.

Q. 49. What are the types of vitamins ?

Ans. According to the solubility, vitamins are divided into two groups :

(1) *Fat soluble vitamins* :

   (a) Vitamin    A
   (b)   ,,        D
   (c)   ,,        E
   (d)   ,,        K

(2) *Water soluble vitamins*

   (a) Vitamin $B$ complex
   (b)   ,,    C
   (c)   ,,    P

Q. 50. What are the sources of vitamins A, D, E and K.

Ans. Sources of vit. A :

*Animal* : Cod-liver oil, milk, butter, eggs fishes.

*Vegetable* : Carrots, spinach, green leaves, mangoes; tomatoes etc.

Sources of vit. D :

*Animal* : Fish-liver oil (Cod-liver oil), butter, milk, eggs.

*Vegetable* : Vegetable negligible.

Sources of vit. E :

*Animal* : Egges, milk, fishes.

*Vegetable* : Seed oils, wheat, soyabean, corn and leafy vegetables.

Sources of vit. K :

*Animal* : Negligible.

*Vegetable* : Cabbage, alfalfa, spinach, tomato, soyabean, etc.

Q. 51. What are the deficiency signs of vitamins A, D, E, and K ?

Ans. *Deficiency signs of vitamin A*

(a) Lack of growth.
(b) Susceptibility to infection.
(c) Night blindness.
(d) Toad skin.
(e) Goose skin.
(f) Degeneration of mucous membrane of cornea and finally the eye-ball.
(g) Ulceration of cornea.
(h) Deformity of bones specially defective deposit of calcium.
(i) Tendency to formation of renal calculi.

*Deficiency signs of vitamin D* :

(a) Reckets (in children).
(b) Osteomalacia (in adult).
(c) Disturbed absorption of calcium salts and its metabolism.

*Deficiency signs of vitamin E* :

(a) Increased degeneration of epithelial tissue in seminiferous tubule leading to oligospermatogenesis.
(b) Decreased secretion of testosterone.

(c) Degeneration of placenta leading premature expulsion of the foetus from the uterus (abortion).

*Deficiency signs of vitamin K :*

Defective blood coagulation leading to haemorrhagic diseases.

**Q. 52.** What are the deficiency signs of vitamins $B_1$, $B_2$, $B_6$, $B_{12}$ and vitamin C ?

**Ans.** *Deficiency signs of vitamin $B_1$ :*

(a) Beriberi.
(b) Polyneuritis or neuritis.
(c) Anorexia.

*Deficiency signs of vitamin $B_2$ :*

(a) Cheilosis.
(b) Angular stomatitis.
(c) Glossitis.
(d) Keratitis.

*Deficiency signs of vitamin $B_6$ :*

(a) Nausea and vomiting.
(b) Diarrhoea without any cause.
(c) Anorexia.
(d) Stomatitis.

*Deficiency signs of vitamin $B_{12}$ :*

(a) Iron deficiency anaemia and all types of anaemia.
(b) Lack of growth.
(c) Deranged erythropoiesis.
(d) Debility.
(e) Degeneration of bone marrow.

*Deficiency signs of vitamin C* :

(a) Decreased immunity.

(b) Scurvey.

(c) Glossitis.

(d) Decreased oxygenation of blood.

(e) Fragility of the epithelial tissue.

Q. 53. What do you mean by normal diet ?

Ans It is one which permits normal growth of the body in all respects.

## EXCRETORY SYSTEM

Q. 54. Name the organs of excretion.

Ans. Kidney, skin, lungs, liver, large intestine and salivary glands.

Q. 55. What are the types of nephron according to the histology ?

Ans. There are two types :

(a) The superficial nephron.
(b) The juxtamedulary nephron.

Nephron is the structural and functional unit of kidney.

Q. 56. Give the functions of kidney.

Ans. (a) It excrets waste products.

(b) It helps to maintain water balance, concentration of blood, osmotic pressure in blood and tissue, regulation of blood pressure.

(c) It plays an important role in the metabolism of vitamin D

Q. 57. Define glycosuria.

Ans. Presence of sugar in the urine is termed as glycosuria.

## ENDOCRINOLOGY

Q. 58. Define hormone.

Ans. Hormone is a chemical messenger secreted by endocrine gland and travels via blood stream, reaches its destination to affect one or more different group of cells.

Q. 59. What are endrocrine glands ?

Ans. There are nine endocrine glands in the human body :

(a) Pituitary gland (Anterior and posterior).
(b) Thyroid gland.
(c) Parathyroid gland.
(d) Pancreas.
(e) Adrenal (Surarenal) glands :
    (i) Adrenal cortex.
    (ii) Adrenal medulla.
(f) Thymus gland.
(g) Pineal body (gland).
(h) Gonads :
    (i) Testis (in male).
    (ii) Ovaries (in female).
(i) Placenta (during pregnancy).

    Pituitary gland is named as "Master gland" also.

**Q. 60.** Name the hormones secreted by different glands.

**Ans. Pituitary glands:**

(a) Growth Hormone (GH) (b) Thyroid stimulating Hormone (TSH) (c) Adrenocorticotrophin (ACT) (d) Lutinising Hormone (LH) (e) Follicle stimulating Hormone (FSH) (f) Prolactin or Luteotrophic Hormone (LTH) (g) Melanocyte-stimulating (MSH) or Intermedin Hormone (h) Vasopressin or Antidiuretic Hormone (ADH) (i) Oxytosin.

**Thyroid Glands:**

(a) Thyroxine.
(b) Tri-iodo-thyronine.
(c) Calcitonin.

**Parathyroid Glands:**

Parathormone (P.T.H).

**Pancreas:**

(a) Insulin.
(b) Glucagon.

**Adrenal (Suprarenal) Glands:**

(a) Glucocorticoids.
(b) Mineralocorticoids. } By Adrenal cortex.
(c) Adrenaline.
(d) Noradrenaline. } By Adrenal medulla.

**Thymus gland:**

Thymosin or Thymin.

**Pineal Body:**

Melatonin.

Gonads

(a) Testosterone or Androgen.
(It is secreted by testis in the case of male).

(b) Oestrogen.
(c) Progesterone.
(d) Androgen.
(e) Relaxin.

} Secreted by ovaries in female.

## Placenta

(a) Human chorionic gonadotrophine (H.C.G).

(b) Chorionic growth Hormone prolactin (C.G.P).

(c) Oestrogen.

(d) Progesterone.

(e) Chorionic thyrotrophin.

(f) Urine relaxin factor (URF).

These hormones are secreted by the placenta in the case of pregnancy.

Q. 61. What do you mean by "Target organ" and "Trophic hormone" ?

Ans. *Target organ.* The organ which is influenced by a particular hormone is called a "Target organ".

*Trophic Hormone.* When the target organ is a endocrine organ itself, the secreted hormone is called Trophic harmone".

All Trophic hormones are secreted by anterior pituitary gland alone.

Q. 62. What do you mean by "Endocrine glands" ?

Ans The glands who have got no duct but they pour their secretion in the blood stream or in lymphatics are known as ductless glands or endocrine glands.

Q. 63. **What is the normal blood sugar level ?**

Ans. It is 80-100 mg. per 100 ml. of blood
— During fasting.
100—120 mg. per. 100 ml. of blood
— After taking meal.

Q. 64. **What do you mean by the following terms :**

(a) Hypoglycaemia.

(b) Hyperglycaemia.

(c) Glycosuria.

Ans. (a) It is a condition in which blood sugar level is present below the normal level.

Such as—below 80 mg. per 100 ml. of blood.

(b) Hyperglycaemia is a condition in which blood sugar increases above the normal level (above 120 mg. per 100 ml. of blood).

(c) *Glycosuria* : Presence of glucose in the urine is known as glycosuria.

In this condition blood glucose level exceeds 180 mg. per 100 ml. of blood whereas the normal blood glucose level is 60—100 mg. per 100. ml. of blood. In this condition the cells of the renal tubule are not able to reabsorb all the glucose. Some quantity of glucose reaches the urinary bladder and glycosuria results.

## REPRODUCTIVE SYSTEM

Q. 65. What do you mean by the following terms :

(a) Puberty (b) Menopause

(c) Menarche    (d) Genes
(e) Heredity.

Ans. (a) *Puberty.* It is a period in male and female when the gonads develop their both endocrine and gametogenic fuctions.

(b) *Menopause.* Ceasation of reproductive capacity after puberty is called menopause. It is usually seen in female in between 45th and 55th years of age. In this period primary and secondary sex organs degenerate

(c) *Menarche.* Onset of first menstruation is named by menarche.

(d) *Genes.* These are D.N.A. particles present in chromosomes, which are carried from one generation to other, carrying the hereditary characters along with it.

(e) *Heredity.* It is the transmission of characters from parent to offspring (According to Dorland's medical dictionary).

Q. 66. What are the functions of testis ?

Ans. (1) Formation of spermatozoa (Spermatogenesis).
(2) Secretion of testosterone.

Q. 67. What is semen ?

Ans. It is a suspension of spermatozoa secreted by the epididymis, prostate, seminal vesicle and cowper's glands.

*Volume.* At each emission it is about 2—4 ml. containing 100—200 million spermatozoa.

Q. 68. Define menstruation ?

Ans. It is a cyclical discharge of blood, mixed with mucus and certain other substances, from the uterus through the vagina in the reproductive life of the females, at an average interval of 28 days
Flow last for 4–6 days without any appreciable pain.

Q. 69. What is the composition of menstruation ?

Ans. The menstruation is composed by :

(a) Blood 30–40 ml.
(b) Stripped of endometrium.
(c) Mucus.
(d) Leucocytes.
(e) Unfertilised ovum.

Q. 70. What are the causes of physiological cessation of menstruation (Ammenorhoea) ?

Ans. The following are the causes :

(a) Before puberty.
(b) After menopause.
(c) During pregnancy.
(d) During lactation.

Q. 71. Define pregnancy.

Ans. Fertilisation of ovum is called pregnancy.

It lasts for 280 days. The end of pregnancy is called parturition.

Q. 72. Define clostrum.

Ans. A secretion of the mammary gland at the end of pregnancy and at the beginning of lactation is known as "Clostrum". It is yellow in colour and secreted for few days after the birth of the child.

**Q. 73.** What are the internal and external reproductive organs of female ?

**Ans.** *Internal organs* :

(a) Uterus
(b) Ovaries
(c) Fallopian tube
(d) Vagina.

*External organs* :

(a) Mons veneris
(b) Labia majora and minora
(c) Clitoris
(d) Hymen
(e) Breast.

**Q. 74.** What are the internal and external reproductive organs of males ?

**Ans.** *Internal organs* :

(a) Testes (two).
(b) Vasdeferens (two).
(c) Seminal vesicles and their ducts.
(d) The prostate gland.

*External organs* :

Penis containing urethra.

**Q. 75.** What are the functions of uterus ?

**Ans.** The following are the functions of uterus :

(a) It receives the fertilized ovum, retains and nourishes the developing foetus throughout the duration of pregnancy.

(b) By the contraction of the muscular walls it expels the foetus at the end of pregnancy.

(c) It acts as an important role in the phenomenon of menstruation.

**Q. 76.** What are the functions of ovary ?

**Ans.** (1) Formation of mature ovum.

(2) Secretion of hormones.

Such as—Oestrogen, progesterone, androgen and relaxin.

(3) It is responsible for all the puberty changes.

(4) It is responsible for the pregnancy and changes associated with it.

**Q. 77.** What are the functions of fallopian tube ?

**Ans.** It collects the ovum discharged from ovary and allows to pass towards the cavity of uterus. Fertilization usually takes place in this tube.

**Q. 78.** Define spermatozoa.

**Ans.** It is the male reproductive cell which develops in the testis and is discharged in the semen through penis having head, neck, body and tail.

**Q. 79.** What are the stages of development of spermatozoa ?

**Ans.** There are five stages of development :
Spermatogonia—Primary Spermatocytic—Secondary Spermatocyte—Spermatid—Spermatozoa.

**Q. 80.** Define spermiogenesis.

**Ans.** It is the sequence of developmental events by which spermatids are converted in mature spermatozoa.

Q. 81. Define semen.

Ans. It is a suspension of spermatozoa secreted by epididymis, prostate gland, seminal vesicle and cower's gland.

It may be also defined as a fluid discharged at ejaculation in the male, consisting of secretion of glands associated with the urogenital tract and containing spermatozoa (Dorland's dictionary).

## NERVOUS SYSTEM

Q. 82. Define neurone

Ans. Neurone is the structural and functional unit of the nervous system.

Q. 83. Define synapse.

Ans. It is the junction where one neurone ends and the other begins.

Q. 84. Define nerve endings.

Ans. These are the terminal bodies where the nerve fibres end in the periphery.

Q. 85. Define reflex action.

Ans. It is an involuntary effector response due to a sensory stimulus.

It is the basic physiological unit of integration in the neural activity.

Q. 86. Define sensations.

Ans. Sensations are feelings aroused by change of environment and carried by sensory nerve to the C.N.S.

Q. 87. Define tracts.

Ans. These are the bundles of fibres carrying one or a group of motor or sensory impulses in the C.N.S.

These are of three types
- (a) Ascending tracts or sensory tracts.
- (b) Descending tracts or motor tracts.
- (c) Intersegmental fibres or ascending and descending both.

Q. 88. What are functions of medulla.

Ans. (1) It controls the cardiovascular functions.
(2) It controls the respiration.

Q. 89. What are the functions of cerebellum ?

Ans. (1) It acts as a receptor for tectile, proprioceptive, auditory and visual impulses.

(2) Each half of the cerebellum controls the tone of the muscles on the same side of the body.

Disturbances in the cerebellum cause the disturbance in (a) posture, (b) voluntary movement and fail to pick up any thing to miscalculation of the distance and speed of movement of the limb, (c) disturbance in standing and (d) disturbance in the tone of the muscle.

Q. 90. What are the functions of cerebrum ?

Ans. (1) It receives all sensory stimuli and convey most of them to consciousness.

(2) It initiates all voluntary movements.

(3) It formulates and associates ideas giving rise to intelligence.

(4) It controls the lower parts of the brain.

(5) It correlates and retains all received impulses, thereby forms the basis of memory.

(6) It exercises unconscious control over many functions of the body.

Q. 91. Define speech.

Ans. Speech is one of the highest faculties of brain and production of articulate sounds, brought about by the co-ordinated activity of different motor, sensory and psychic centres which always bears a definite meaning.

# PART III

# Pathology Oral Question and Answer

CHAPTER 1

# TERMINOLOGY

Q. 1. Define Pathology.

Ans. Pathology is a branch of medical science which deals with the study of the abnormal functions (diseases) of the different parts of the body.

Q. 2. Define the following terms :
   (a) General Pathology.     (b) Clinical Pathology.
   (c) Special Pathology.     (d) Laboratory Pathology.
   (e) Postmortem Pathology.
   (f) Experimental Pathology. (g) Gross Pathology.
   (h) Aetiology.   (i) Morphology.   (j) Histology
   (k) Bacteriology.

Ans. (a) *General Pathology.* It is a branch of pathology which deals with the principles, theories, explanations and classifications of diseases.

   (b) *Clinical Pathology.* It deals with the signs and symptoms, stages, course and diagnosis of the disease of a patient.

(c) *Special Pathology.* It is a branch of pathology which deals with the method to find out the course, specific nature and decision regarding medical or surgical treatment of the diseases.

       Such as—Study of syphilis, Tuberclosis, Cancer etc. comes under special pathology.

(d) *Laboratory Pathology.* It is that branch of pathology which deals with the study of diseases for the purpose of diagnosis by laboratory examination of blood, urine, stool, sputum etc.

(e) *Postmortem Pathology.* The study of the disease after death of the patient is called postmortem pathology.

(f) *Experimental Pathology.* This includes the study of the disease under artificial condition.

       Such as—Study of Cancer regarding origin and developement on animal.

(g) *Gross Pathology.* It is a branch of pathology by which we know the structural changes by the help of naked eye.

(h) *Aetiology.* It is the study of the cause of the disease.

(i) *Morphology.* The study of the shape and size of the organs or the micro-organism by microscope is called morphology.

(j) *Histology.* It deals with the study of structure and structural changes by the help of microscope.

(k) *Bacteriology.* The study of bacteria is called bacteriology.

**Q. 3.** Define the following terms :

(a) Pathogenic Bacteria.
(b) Non-pathogenic Bacteria.
(c) Facultative Parasites.   (d) Bacteria.

**Ans.** (a) *Pathogenic Bacteria.* Those micro-organism which can produce diseases are termed as pathogenic bacteria or pathogenic micro-organism.

(b) *Non-pathogenic Bacteria.* Those types of bacteria which reside in the body but they can not produce any kind of disease.

(c) *Facultative Parasites.* These are non-pathogenic. They reside in the body but under some circumstances they become pathogenic.

Such as—B. coli.

(d) *Bacteria.* These are unicellular plant like micro-organism who have got no chlorophyll.

**Q. 4** What do you mean by Gram Positive and Gram Negative Bacteria ?

**Ans.** Those types of bacteria which can be stained by gram stain are named as Gram positive bacteria. Those bacteria which can not stained by Gram stain are called Gram negative bacteria.

Such as—*Gram Positive*—Streptococci, Staphylococci, Tuberote bacilli, Bacilus leprae, B. Tetanus, B. Anthrax, B. Diphtheriae, etc.

*Gram Negative.* Pneumococcus, Meningococcus, Gonococcus, N. Catarrhalis, B. Pestis, B. Influenzae, B. Coli, B. Dysenteriae, B. Typhosum, B. Paratyphosus, V. Comma, Spirochaeta Pallida.

Q. 5. Define the following terms :
- (a) Host
- (b) Parasitism
- (c) Symboisis
- (d) Obligatory Parasite
- (e) Facultative Parasite
- (f) Parasite
- (g) Parasitology
- (h) Protozoology
- (i) Helminthology
- (j) Entomology
- (k) Entozoa
- (l) Ectoparasite
- (m) Endoparasite
- (n) Commensal
- (o) Infection
- (p) Infestation
- (q) Auto-infection
- (r) Super infection.

Ans. (a) *Host.* Host is the organism which harbour the parasite.

(b) *Parasitism.* Parasitism is the relationship between the host and guest (Parasite).

(c) *Symboisis.* It is the definite association of the two specifically different organism. So their life is dependent upon each other. They can not live without each other.

(d) *Obligatory Parasite.* Obligatory parasite is that organism whose existence entirely depends upon the host.

(e) *Facultative Parasite.* The parasite which is capable of independent existence together with the parasitic life is known as facultative parasite.

(f) *Parasite.* Parasites are the organism which depend their life at the cost of host.

(g) *Parasitology.* Parasitology is that branch of science which deals with the study of parasites and their pathogenecity to the host.

F7B

(h) *Protozoology.* It is that branch of science which deals with the study of protozoa.

(i) *Helminthology.* It deals with the study of Helminth or Metazoa.

(j) *Entomology.* Entomology is that branch of science which deals with the study of Anthropoid or Anthropoidea.

(k) *Entozoa.* $Ent = Inside$, $Zoa = Life\ of\ Parasite$. Entozoa is that branch of science which deals with the study of life of parasite inside the body tissues.

(l) *Ectoparasite.* It is that type of parasite which lives in the outer surface of the body.

(m) *Endoparasite.* Endoparasite is that type of parasite which lives in the inner side of the body.

(n) *Commensal.* It means that organism or parasite which are benefited from host but they neither benefit the host nor harm the host.

(o) *Infection.* Infection is the pathological condition in which the organism enters the body, multiply in the body and cause disease to the host.

According to "Manikoff" the infection is nothing but the struggle between the host and guest.

(p) *Infestation.* Infestation is the pathological condition in which the macroparasite that is Helminth and Anthropods establish on the superficial or the surface tissue of the body.

(q) *Auto-infection.* It means the host is the direct source of infection.

Such as—Thread worm (Entrobilous Vernicularis) which cause auto-infection from the anal cannal to the mouth via finger.

(r) *Super infection.* When the host harbouring the parasite or the organism and again re-infection takes place by the same organism is known as super-infection or re-infection.

## CHAPTER 2

# NECROSIS AND GANGRENE

Q. 1. Define Necrosis ?

Ans. Local death of tissue of the living body is named as necrosis which occurs suddenly.

Q. 2. Define Necrobiosis ?

Ans. It is a term applied for slow death of individual cell due to infilteration of abnormal constitnents or degeneration of the contents of the cells.

Q. 3. What are the varieties of necrosis ?

Ans. There are six varieties of Necrosis :
 (a) Toxic necrosis.
 (b) Fat necrosis.
 (c) Traumatic necrosis.
 (d) Coagulation necrosis.
 (e) Caseation necrosis.
 (f) Liquefaction necrosis.

Q. 4. What are the causes of Necrosis ?

Ans. (i) Physical cause : such as—Burn, injury, extreme cold, heat stroke etc.

(ii) Loss of supply of blood due to any cause, such as—thrombosis, embolism. obstruction in the blood vessels (arteries).

(iii) Micro-organism.

(iv) Chemical agents.

**Q. 5.** Define Gangrene.

**Ans.** Death of tissue (Necrosis) with putrefaction is termed as gangrene.

**Q. 6.** What are the types of Gangrene ?

**Ans.** These are five :

(a) Moist gangrene.

(b) Dry gangrene.

(c) Gas gangrene.

(d) Diabetic gangrene.

(e) Senile gangrene.

## INFECTIONS

**Q. 7.** Define Infection.

**Ans.** It is the invasion of tissue of the living body by pathogenic micro-organism which then multiply, liberate toxins and cause disease.

**Q. 8.** What are the mode of infection ?

**Ans.** Mode of infection :

(a) Infection takes place by polluted food and drinks such as—Cholera, dysentery, diarrhoea, infective hepatitis, typhoid, paratyphoid, etc.

(b) *By droplet.* Pneumonea, diphtheria, whooping cough, influenza, pox, tuberculosis, etc.

(c) *By contact.* Gonorrhoea, syphilis, etc.

(d) *By injection and blood transfusion*—Syphilis, serum sickness, etc.

(e) *By insects.* Such as—Mosquitos, flies, fleas, lice, etc.

Q. 9. What are the types of infection?

Ans. There are following types of Infection:

(1) *Primary infection.* This is the first noted in an illness.

(2) *Secondary infection.* In weakned body, by the primary infection, in many cases secondary infection takes place.

Such as—Otitis in typhoid, diarrhoea or cough after measles.

(3) *Mixed infection.* In this case the disease is caused by more than one organism.

(4) *Local infection.* This infection takes place in a localised area.

Such as—Wound or abscess.

(5) *Focal infection.* In this infection the organisms are confined to one area but may cause further infection to the other part.

(6) *General or systemic infection.* In this infection entire body seems to be affected and this type of infection may be classified as:

Bacteraemia, Septicaemia, Toxaemia, Sapraemia.

Q. 10. What are the mode of spread of Infection ?

Ans. This takes place in the following ways :

(1) *Epidemic form.* In a locality many persons suffering from same infection.

(2) *Endemic form.* Many persons suffering from the prevailing disease in a particular region.

Such as—Goitre in certain Himalayan region.

Scrotal hydrocle in certain area, etc.

(3) *Sporadic form.* Epidemic of the same disease found in here and there.

(4) *Pandemic form.* Epidemic of a disease occurring to a very wide area or the whole world.

## INFLAMMATION

Q. 11. Define Inflammation ?

Ans. Inflammation is a protective process to injury or irritation by local tissue by which cells and exudade accumulate in the injuired tissue and tend to protect them from further injury.

Q. 12. What are the fundamental characters of inflammation ?

Ans. There are five fundamental characters of inflammation.

(1) *Congestion* (*Rubor*). It occurs due to increased supply of blood to the affected part.

(2) *Swelling* (*Tumor*). It occurs due to accumulation of blood serum and cells to the affected part.

(3) *Pain* (*Dolar*). It is due to irritation of sensory nerve endings of the part.

(4) *Heat* (*Calor*). Heat is due to accumulation of blood.

(5) *Loss of Function*. It is due to ill-condition of the parenchymatous cells (functioning cells).

**Q. 13.** What are the classifications of Inflammation?

**Ans.** There are three classes of inflammation :

  (a) Acute form.
  (b) Sub-acute form.
  (c) Chronic form.

**Q. 14.** What are the pathological changes during Inflammation?

**Ans.** There are two types of pathological changes (Pathology) of inflammation :

### (1) Tissue changes

(a) In the tissue plasma of blood accumulates and dilutes the toxins of the pathogenic organism and try to remove its harmful effect.

(b) W.B.C. of blood eat up the bacteria.

(c) The plasma helps in the healing of wound by the formation of fibria.

(d) In chronic inflammation neutrophils of W.B.C. help in the destruction of bacteria.

### (2) Circulatory changes

(a) The blood vessels of the affected part dilate due to nerve irritation and the flow of blood is accelerated.

(b) When the blood vessel is full with blood, the blood flow stops and the leakage of plasma from the pores of the wall of the blood vessel into the tissue takes place.

(c) W.B.C. of blood also migrate to the infected part and engulf the bacteria. Later on in the case of chronic inflammation or in the case of tuberculosis, leprosy, etc. the giant cells also migrate from the blood vessels into the tissue and try to destroy them.

Q. 15. What are the sequelae of inflammatory reaction?

Ans. Sequelae are the following :

    (a) Hypertrophy     (b) Atrophy
    (c) Ulcer formation     (d) Gangrene
    (e) Necrosis     (f) Degeneration
    (g) Formation of keloid   (h) Formation of cancer.

## IMMUNITY

Q. 16. What do you mean by Immunity?

Ans. Immunity is a resistance power of the living body to protect the infection or infectious disease.

Q. 17. What are the varieties of Immunity?

Ans. It has got two varieties as follow :

    (1) Natural Immunity.

    (2) Artificial or Acquired Immunity.

    (1) **Natural Immunity**

        With this immunity an individual is born. It is of four types :

(a) *Individual Immunity.* It is a characteristic of a particular person.

Such as—In the case of epidemic form of cholera a person may not be affected by cholera.

(b) *Racial Immunity.* This type of natural immunity is found in a particular race within a species.

Such as—Negroes are immune to yellow fever.

(c) *Congenital Immunity.* Due to passive transfer of antibodies from mother to child through the placenta, the child immunised to prevent some diseases. So, this type of immunity is found in the new-born.

Such as—The infant in the first year of life has got immunity to prevent scarlet fever and diphtheria.

(d) *Species Immunity.* This type of immunity is found in a particular species.

Such as—Dogs are resistant to tuberculosis

## (2) Artificial Immunity

This type of immunity is acquired by the living body during the life-time. It has got the following classification :

(A) Active acquired immunity.

(B) Passive acquired immunity.

(A) *Active acquired immunity.* It is that type of immunity in which an individual immunises due to development of antibodies within the individual by coming in the contact with the microorganism or their products. The cells and tissues of the body themselves react to produce the specific immunity.

It is of two types ;

(a) *Artificial active immunity.* It is acquired by a person from a course of specific vaccination.

Such as—Cholera vaccination, small pox vaccination, etc.

(b) *Natural acquired active immunity.* This type of immunity is a result of an attack of the disease. One attack of certain disease confers life long immunity.

For example—Whooping cough, diphtheria, typhoid, scarlet fever, etc.

(B) *Passive acquired immunity.* This is a temporary acquired immunity in which antibodies are formed in another animal and the blood serum of that very animal is injected into the person. In this type of immunity the cells and tissues of the body of the vaccinated person take no part in producing the immunity.

For example—A.T.S. for Tetanus.

Immune Gamma—Globulin for poliomyetitis, etc.

Q. 18. What do you mean by Antigen and Antibody ?

Ans. **Antigen.** It is a type of protein which has got a power to stimulate the production of specific antibodies.

Specificity of the antigen is known by the study of its chemical composition.

**Antibody.** In the response to the presence of antigen an antibody is formed by the animal body.

## CIRCULATORY DISTURBANCES

### (1) Embolism

Q. 19. Define the term Embolism.

Ans. It is a term applied for the obstruction in the blood circulation due to presence of foreign body in the blood vessel.

Q. 20. What do you mean by Embolus ?

Ans. The foreign body when obstructs the blood flow by entering the blood vessel is known as "Embolus".

### (2) Thrombosis

Q. 21. What do you mean by thrombosis ?

Ans. When the clotting of blood occurs inside the blood vessel during life and the particle of clot when obstructs the blood flow then that condition is termed as Thrombosis.

Q. 22. What do you mean by Thrombus ?

Ans. The particle of blood clot, inside the blood vessel, when obstructs the flow of blood is named as "Thrombus".

Q. 23. What is the composition of Thrombus ?

Ans. The thrombus is composed by the following :

(a) W.B.Cs.

- (b) R.B.Cs.
- (c) Platlets.
- (d) Fibrin.

Q. 24. What are the types of Thrombosis :

Ans. There are two types :
- (a) *Arterial Thrombosis.* Obstruction in the arterial flow.
- (b) *Venous Thrombosis.* Obstruction in the venous flow.

## (3) Hyperaemia or Congestion

Q. 25. Define Hyperaemia.

Ans. It is a condition in which the blood accumulation takes place in the blood vessels of the part of the body due to this there is disturbance in the distribution of blood.

Q. 26. What are the types of Hyperaemia ?

Ans. It is of two types :
- (a) *Active Hyperaemia.* Accumulation of blood in artery.
- (b) *Passive Hyperaemia.* Accumulation of blood in the veins.

Passive Hyperaeima is of two types :
(i) Local            (ii) General.

## (4) Haemorrhage or Bleeding

Q. 27. What do you mean by Haemorrhage ?

Ans. It is a condition in which blood escapes from blood vessels.

Q. 28. What are the types of Haemorrhage ?

Ans. It is of two types :

- (a) *Arterial or Active Haemorrhage.* Bleeding takes place from ruptured artery.
- († ) *Venous Haemorrhage.* Bleeding occurs from ruptured veins.

**Note.**
- (a) *Capillary Haemorrhage.* It is the oozing of blood from capillaries. It is internal haemorrhage.
- (b) *Petechiae.* It is also a capillary haemorrhage in the skin or serous membranes.
- (c) *Ecchymosis.* It is the extra vasation of blood in the skin due to blunt injury.
- (d) *Cerebral Haemorrhage.* Bleeding from brain.
- (e) *Epistaxis.* Bleeding from nose.
- (f) *Haemoptysis.* Bleeding from lungs.
- (g) *Haematemesis.* Bleeding from stomach.
- (h) *Malaena.* Bleeding from intestine.
- (i) *Haematuria.* Bleeding from urinary tract.
- (j) *Haemorrhagia.* Excessive bleeding during menstruation.
- (k) *Haemorrhoids.* Bleeding from piles.
- (l) *Haemophilia.* Bleeding from slightest injury and does not stop without proper treatment. It is hereditary disease.
- (m) *Haematocele.* Bleeding in the Tunica vaginalis of testis.

**Q. 29.** What are the difference between arterial and venous haemorrhage?

**Ans.** Difference between arterial and venous haemorrhage:

| *Arterial Haemorrhage* | *Venous Haemorrhage* |
|---|---|
| (1) Blood is bright red | Blood is dark red. |
| (2) It comes out in gushes and does not stop unless pressure is applied. | It comes out slowly and may stop without interference. |
| (3) Causes great weakness. | Weakness not marked. |
| (4) It may become fatal if it is not stopped. | It is not fetal unless very large quantity of blood has gone out. |
| (5) Epistaxis is example. | Bleeding from piles is example. |

### (5) Cyanosis

**Q. 30.** Define the term Cyanosis.

**Ans.** It is a condition in which venous blood accumulates in the skin and marked by blueness of the parts. It occurs due to insufficient supply of oxygen which causes changes in the haemoglobin (Hb).

### (6) Anoxaemia

Q. 31. Define Anoxaemia.

Ans. Lack of oxygen in the blood is known as Anoxaemia.

### (7) Shock

Q. 32. What do you mean by Shock?

Ans. It is a depressed condition of body function or vitality characterized basically by reduction in the effectively circulating blood and in blood pressure (circulatory collapse) and caused by injuries or other conditions.

Q. 33. What are the symptoms of shock ?

Ans. The following are the symptoms of shock :

    (a) Gidiness, faintness, shallow and rapid respiration.

    (b) Temperature becomes sub-normal (Below 97°F.).

    (c) Face pale, nausea and vomiting may occur.

    (d) Profuse sweating and coldness of body.

    (e) Blood pressure falls, puls becomes weak and rate also falls.

    (f) Cyanosis of lips.

    (g) Unconsciousness, comatose in severe case.

    (h) Patient may die of heart failure.

## DISORDERS OF BLOOD

Q. 34. What are the kinds of diseases of R.B.C. ?

Ans. There are two kinds of diseases of R.B.C. :

    (a) *Anaemia.* In this condition total number of R.B.C. falls and haemoglobin reduced.

    (b) *Polycythaemia.* In this condition total number of R.B.C. increases above the normal.

The total number of R.B.C. is as follows :

(a) In adult male    —5 million/cu. mm. of blood.
(b) In adult female  —4·5 ,,  ,,  ,,
(c) In infant          —6·7 ,,  ,,  ,,
(d) In foetus         —7·8 ,,  ,,  ,,

**Q. 35.** What are the classifications of anaemia caused by R.B.C. disorder?

**Ans.** Anaemia of R.B.C. disorder is classified as follow:

## (A) Aetiological

  (a) *Haemolytic Anaemia*: It is caused by excessive destruction of R.B.C.

  (b) *Haemorrhagic Anaemia*: It is caused by excessive bleeding.

  (c) *Dyshaemopoietic Anaemia*. It is due to decreased formation of blood due to disease of Haemopoitic system.

## (B) Morphological.

This type of anaemia is caused by disturbed morphology of the red blood cells.

It has got following types:

(a) *Microcytic Anaemia*:

In this condition size of the R.B.C. decreases but haemoglobin content increases.

(b) *Hypochromic Microcytic Anaemia*:

In this condition haemoglobin and size of the R.B.C. both decrease or diminish.

(c) *Macrocytic Anaemia*:

The R.B.Cs. become larger than the normal but haemoglobin reduces.

(d) *Normocytic Anaemia*:

The total number of R.B.C. reduced only.

## DISEASES OF W.B.C.

**Q. 36.** What are the diseases caused by W.B.C. disorders?

**Ans.** Following diseases are caused by W.B.C. disorders :

(a) Leukaemia.
(b) Leukopenia.
(c) Leukocytosis.

**Q. 37.** What do you mean by Leukaemia?

**Ans.** It is a disease characterised by an abnormal and excessive proliferation of leucopoitic tissues of the body with abnormally increased total number of circulating immature forms of W.B.Cs.

It is a fatal disease having acute and chronic varieties. The cause of this disease is not known till now. It is also named as *"Blood Cancer"*.

**Q. 38.** What do you mean by Leukopenia?

**Ans.** It is that condition in which total count of W.B.Cs. falls than the normal.

**Q. 39.** What do you mean by Leukocytosis?

It is a condition in which number of W.B.Cs. increase.

It is of five types as follow :

(a) Neutrophilic Leukocytosis
(b) Eosinophilic ,,
(c) Basophilic ,,
(d) Lymphocytic ,,
(e) Monocytic ,,

## DISEASES OF PLATELETS

**Q. 40.** What are the diseases produced by platelets?

Ans. Diseases are of two types of Platelets :

(a) *Thrombocytosis.* Total number of Platelets are abnormally increased.

(b) *Thrombocytopenia.* Total number of Platelets are abnormally decreased.

The normal total count of Platelets—2, 50000—4, 50000 per cu. mm. of blood.

## TUMOUR OR NEOPLASM

Q. 41. Define Tumour.

Ans. It is a condition of pathological swelling or enlargement or overgrowth of tissue which is progressive, parasitic and often disruptive cell proliferation in nature.

It is also known as New-growth.

Q. 42. What are the varieties of Tumour or Neoplasm ?

Ans. The tumours are divided into three classes as follow:

### (A) Histoma or Connective tissue tumour

It is of two types :

(a) Benign (Simple).

(b) Malignant (Sarcoma).

### (B) Cytoma or Epithelial tissue tumour

(a) Benign (Simple).

(b) Malignant (Carcinoma).

### (C) Teratoma or Mixed cell tumour

They are usually malignant type.

Q. 43. What are the difference between Benign and Malignant tumours?

Ans. Following are the differenciating points of Benign and Malignant tumours:

| *Benign Tumour* | *Malignant Tumour* |
|---|---|
| (1) Growth is slow | Growth is rapid |
| (2) Encapsulated | No tendency for encapsulation. |
| (3) No tendency for infilteration | It infilterates into the neighbouring tissue. |
| (4) Tendency to disorganise the tissues is not found. | Tendency to disorganise the tissue is present and it can form open ulcer. |
| (5) Metastasis absent | Metastasis present. |
| (6) Bleeding and ulceration is not common. | Bleeding and ulceration is very common. |
| (7) In the shape and size, cells are similar to the tissue and uniform in size. | Shape and size tend to vary. |
| (8) Supply of blood vessels is few and well developed. | Many ill-developed and malignant cell's lined blood vessels are present. |
| (9) Stroma cells are in big bands. | Fine and reticulate. |
| (10) Nuclei are matured. | Multiple and imatured cell like. |

(11) No tendency to recur after excision.     It recur after excision.

(12) No harmful to host     Harmful to host.

(13) Stain of nuclei is of normal colour.     Hyperchromatic nuclei.

**Q. 44.** What is the difference between tumour and cyst?

Ans.

| Tumour | Cyst |
|---|---|
| (1) It is solid. | It is filled with fluid. |
| (2) It is not yield to pressure. | It yields to pressure. |

**Q. 45.** What is the difference between tumour and oedema.

Ans.

| Tumour | Oedema |
|---|---|
| (1) Tumour does not pit on pressure. | Oedema pits on pressure. |
| (2) Tumour is localised. | Oedema may occur over large part of the body. |

**Q. 46.** What is the difference between tumour and inflammation?

Ans.

| Tumour | Inflammation |
|---|---|
| Inflammatory symptoms are not present in tumours except swelling. | Inflammatory symptoms are present in inflammation with swelling. |
| | Such as—Redness, heat, pain, and loss of function. |

**Q. 47.** Define Carcinoma.

Ans. Carcinoma is a malignant tumour made up of connective tissue enclosing epithelial cells.

Q. 48. What are the varities of Carchinoma ?

Ans. Varities :

   (i) Epithelioma.    (ii) Columnar celled carcinoma.
   (iii) Glandular celled carcinoma.
   (iv) Chorionic epithelioma.

## OEDEMA

Q. 49. Define Oedema.

Ans. Collection of fluid in the cavity of tissue is named as oedema.

Q. 50. What are the characters of Oedema ?

Ans. Characters of Oedema :

   (a) It pits on pressure.
   (b) It is not red and painful as inflammation.
   (c) A thin watery and non-coagulable fluid oozes from puncture ; if it is punctured.
   (d) The fluid does not contain bacteria, pus blood but it contains small quantity of proteins.

Q. 51. What are the causes of Oedema ?

Ans. Oedema may caused by the following factors :

   (1) Cardiac failure     (2) Cirrhosis of liver
   (3) Allergic            (4) Beri-beri
   (5) Filarial lamphagitis (6) Anaemia
   (7) Renal disease       (8) Puerperal thrombophlebitis
   (9) Tuberculosis of peritoneum.

Q. 52. What are the clinical varieties of Oedema ?

Ans. The following are the varieties of oedema :
- (a) *Anasarca*. Oedema of whole body.
- (b) *Myxoedema*. It is solid oedema of hypothyroidism.
- (c) *Dropsy*. Same meaning as oedema.
- (d) *Angio—Neurotic oedema*. It is sudden oedema of short duration of whole body.
- (e) *Ascites*. Collection of serous fluid in the cavity of peritoneum.

## ALLERGY

Q. 53. Define the following terms :

- (a) Allergy
- (b) Susceptibility
- (c) Idiosyncrasy
- (d) Irritability
- (e) Excitability
- (f) Anaphylaxis
- (g) Sensitization
- (h) Elective affinity
- (i) Desensitization
- (j) Atopy.

Ans. (a) *Allergy*. It is an abnormal state characterised by hypersensitivity to a physical or chemical agent which shows itself by immediate influence and abnormal response.

For example—Eating of guava causes coryza in a person or in some persons out of so many persons.

(b) *Susceptibility*. It is the lack of resistance of a body to the effects of a beneficial or a deleterious agent.

It is the quantity of the mind and body for an influence.

(c) *Idiosyncracy.* It is a state of hypersensitivity and hypersusceptibility peculiar to an individual as a whole whereas allergy belongs to the cells and organs of the body, idiosyncrasy belongs to the whole organism.

(d) *Irritability.* It is a nervous or tissue response of abnormal responsiveness to slight stimuli.

(e) *Excitability.* Readiness to respond to a stimulus is named as excitability.

It is a nervous response present mostly in allergic persons.

(f) *Anaphylaxis.* It is a exaggerated reaction of a person to a foreign protein or other substance to which he has previously become sensitized.

(g) *Sensitization.* It is a process of becoming sensitive.

(h) *Elective affinity.* It is a relation between chemical or biochemical substances to cells, tissues and organs of the body.

(i) *Desensitization.* It is a process of abolition of sensitivity to a particular antigen.

(j) *Atopy.* It is a hypersensitivity due to hereditary influences.

## EXUDATION AND SUPPURATION

Q. 54. Define Exudation.

Ans. Inflammatory discharge is called **exudation**. It may be thin, thick, blend, muco-purulent, stingy, ropy or false membranous.

Q. 55. Define Suppuration.

Ans. Formation of pus due to infection is named as suppuration.

Q. 56. Define Catarrh.

Ans. Discharge of mucous from congested mucous membrane due to allergic condition is called catarrh.

Q. 57. What are the causes of suppuration?

Ans. It is produced by the infection of pyogenic bacteria.

Q. 58. What are the factors on which different colour of pus depend?

Ans. The different colour of pus depend upon the infection of different types of bacteria.

Such as:

| Name of Bacteria | Colour of Pus |
| --- | --- |
| (1) Meningococcus | Thick white pus |
| (2) Pneumococcus | Yellowish white pus |
| (3) Staphylococcus albus | White pus |
| (4) Staphylococcus citreous | Lemon yellow pus |
| (5) Staphylococcus aureous | Golden yellow pus |
| (6) Straptolococcus. haemolyticus | White and yellow pus. |
| (7) Gonococcus | Yellowish white pus |
| (8) B. Coli | Whitish pus |
| (9) Micrococcus catarahalis | White and stringy pus |
| (10) B. Pestis | Yellowish white pus |
| (11) B. Dysentery | Blood and white pus |
| (12) B. Pyocyneous | Bluish pus. |

## FEVER OR PYREXIA

Q. 59. Define Fever or Pyrexia.

Ans. Rise of body temperature between 99°F. and 105°F. and onward is called fever or pyrexia.

Q. 60. Define Hyperpyrexia.

Ans. Rise of body temperature above 107°F. is called hyperpyrexia.

Q. 61. Why body temperature rises?

Ans. The temperature of body rises due to derrangement of the heat regulating mechanism. The toxins (Pyrogens) acts on W.B.C. and form endogenous pyrogen. This endogenous pyrogen acts directly on the anterior hypothalamus and due to this the body temperature rises.

Q. 62. What are the causes of fever ?

Ans. (1) Fever occurs due to infection of pathogenic microorganism.

(2) Due to poisons and certain drugs.

(3) Due to injuries.

(4) Due to allergic diseases.

Such as—Menstrual fever, catarrhal fever, urticarial fever, etc.

(5) Due to malignant diseases.

(6) Due to endocrine and metabolic diseases.

(7) Pyrexia of unknown origin (P.U.O.).

Q. 62. What is the normal body temperature ?

Ans. The human body temperature in health is 98.4°F. or 37°C.

Q. 64. What do you mean by Apyrexia ?

Ans. It is that condition in which the body temperature is normal.

Q. 65. What are the physiological responses due to fever ?

Ans. Following are the physiological responses due to fever :

(a) Blood pressure increases.

(b) Puls rate increases.

(c) Cardiac output increases.

(d) Negative nitrogen balance occurs.

(e) Dehydration and fall of plasma chloride level occurs.

(f) Metabolism increases.

(g) Respiration rate increases.

(h) Carbohydrate and fat metabolism disturbed leading to increased production of acid in the body and urinary ammonia also increases.

Q. 66 Give the name of the instrument by which temperature is measured ?

Ans. Clinical Thermometer.

Q. 67. Give the name of the parts of the body from where temperature is taken.

Ans. Temperature is taken from :

(a) Axilla

(b) Mouth

(c) Rectum.

The temperature taken from axilla may not very correct. It is usually one degree less the mouth temperature. Rectum temperature is the correct temperature. It may be one degree higher than the oral temperature in certain conditions.

Such as—In cholera, the rectal temperature is one degree higher than the oral.

## REGENERATION

Q. 68. Give the definition of Regeneration.

Ans. Regeneration is a process by which new tissue is similar to destroyed one at the site of destroyed tissue.

Such as—Formation of blood after Haemorrhage (bleeding).

Q. 69. What are the kinds of Regeneration ?

Ans. Regeneration has got two kinds :

(a) *Physiological Regeneration.* It is a process by which replacement of tissues takes place for those who are destroyed during physiological activity of the body.

(b) *Pathological Regeneration.* This replacement takes place after the destruction of tissue due to diseased conditions or after injury.

## DEGENERATION

Q. 70. Give the definition of Degeneration.

Ans. It is a process by which deterioration of functioning tissue (Parenchymatous tissue) of an organ takes place during the life-time.

**Q. 71. What are the types of Degeneration?**

Ans. Degenerations are of following types:

### (A) Albumen Degenerotion (Cloudy Swelling)

In this type, the process of degeneration often occurs in those organs or tissues who are rich in paranchymatous cells.

It is very mildest type of degeneration.

Such as—Kidney, heart, liver, etc.

### (B) Fatty Degeneration

It takes place in the paranchymatous cells and characterised by appearance of minute droplets of facts in the cytoplasm, *e.g.* Liver, heart, kidneys, etc.

### (C) Hyaline Degeneration

Degeneration of the cells take place due to deposition of a dense firm and translucent homogenous substance which is of protein nature.

The tissue or cells commonly affected by this process are tubules of the kidney, epithelial cells, etc.

According to the nature of deposited hyaline matter in the tissue during degenerative process, hyaline degeneration is classified in the following classes:

(a) *Amyloid Degeneration.* Accumulation of amyloid matter (firm, waxy and semitranslucent albuminous material) takes place.

It occurs generally in spleen and liver.

(b) *Colloid Degeneration.* Accumulation of colloidal material (firm, transparent and glue like substance) takes place in this type of degeneration.

Such as—Colloidal renal cysts, colloidal overian tumours, colloidal goitre, etc.

(c) *Mucoid Degeneration.* Accumlation of mucin (a slimy substance) takes place.

It takes place generally respiratory and digestive organs.

(d) *Zenker's Degeneration.* This type of degeneration takes place in the muscles due to high temperature in typhoid and toxic conditions.

Q. 72. Give the definition of Pyknosis.

Ans. It is nothing but only thickening and shrinkage of the nucleus after death of a cell.

Q. 73. Give the definition of Karyolysis.

Ans. It is an engymatic action on the cell by which cell liquify and cytoplasm and mucleus of the cell destroy.

These enzymes are amylolytic, lipolytic and proteolytic.

## INFILTRATION

Q. 74. Give the definition of Infiltration.

Ans. Infiltration is the deposition or diffusion of a morbid liquid or solid in the cells and tissues of the living organism.

**Q. 75.** What are the types of infiltration ?

**Ans.** There are nine types of infiltration :

(a) *Adipose or fatty infiltration* :

It is abnormal deposition of fat within healthy cells as the result of a systemic metabolic derangement.

(b) *Calcarious infiltration* :

In this type deposition of calcium and earthy salts in the tissue takes place.

(c) *Cellular infiltration* :

It is the infiltration of tissue with round cells.

(d) *Glycogenic infiltration* :

It is the deposit of glycogen in the cells or tissue.

(e) *Pigmentary infiltration* :

In this type of infiltration accumulation of pigment in the tissue takes place.

(f) *Purulent infiltration* :

Presence of deposited pus cells in a tissue is named as purulent infiltration.

(g) *Serous infiltration* :

Abnormal accumulation of serum in the tissue is called serous infiltration.

(h) *Urinous infiltration* :

The excessive or profuse collection of urine in a tissue is called urinous infiltration.

(i) *Waxy infiltration* :

It is the deposition of amyloid substance in the cells or tissue.

## ATROPHY

Q. 76. Give the definition of Atrophy.

Ans. The decreasment in shape and size of a normally developed organ or tissue is called atrophy.

Q. 77. What are the causes of Atrophy ?

Ans. Following are causes of atrophy :

(1) Physiological causes

It occurs due to non-use of the organ.

Such as—Atrophy of thymus gland and appendix in mankind.

(2) Congenital cause.

(3) Endocrine disturbances.

(4) Senile atrophy. Such as—Atrophy of uterus and breast in old female.

(5) Loss of blood supply to the part.

(6) Atrophy due to increased pressure on the part.

(7) Atrophy due to tonic affect.

(8) Abnormal supply of nutrition.

Q. 78. What do you mean by Marasmus ?

Ans. It is a condition specially found in the young infants characterised by progressive wasting.

It is also named as "Infantile atrophy"

Q. 79. What is Aplasia ?

Ans. It is a condition characterised by failing of an organ to develop.

Q. 80. What is Kwashiorkor ?

Ans. It is a syndrome occurs due to severe nutritional deficiency.

Q. 81. What is Hypoplasia ?

Ans. It is a condition marked by stopped growth of the organ.

## HYPERTROPHY

Q. 82. Define Hypertrophy.

Ans. Enlargement of a part or more than one part of the body or organ with corresponding functional increasment is called hypertrophy.

Q. 83. What are the causes of Hypertrophy ?

Ans. Generally hypertrophy occurs due to following causes.

(1) Hypertrophy caused by physiological causes.

*For Example* :— Hypertrophy occurs due to increased work of the part or organ. Due to change in the normal function of the body. Such :—
Hypertrophy of mammary gland during pregnancy
Hypertrophy also occurs when a part of the body is doing double work in respect of the other part. This is called compensatory hypertrophy.

(2) Pathological causes of hypertrophy.

Such as—Gigantism, pseudo-hypertrophy, fatness etc.

Q. 84. What do you mean by the following terms?

(a) Hetroplasia

(b) Hyperplasia

(c) Mataplasia

(d) Anaplasia.

Ans. (a) *Hetroplasia* : It is a condition in which growth of abnormal tissue at the place of normal tissue occurs.

Such as—At the place of scar, formation of keloid takes place.

(b) *Hyperplasia* :—It is condition of increased volume of a tissue or organ caused by formation and growth of new cells or tissue or connective tissue after injury or inflammation.

(c) *Mataplasia* :—In this condition transformation of one type of tissue into another takes place.

(d) *Anaplasia* :—It is a condition of irreversible alteration in adult cells into embryonic cell types. Such as—Malignant tumour cells and carcinomatous cells.

## REPARATIVE PROCESS

Q. 85. Define Reparative Process.

Ans. It is a process by which healing takes place after destruction of any tissue due to any cause.

Q. 86. What are the factors on which healing depends?

Ans. Healing depends on the following :

    (i) Rest to the part which is affected.

    (ii) Separated segments should be attached.

    (iii) Removal of foreign body.

    (iv) Aseptic measure.

    (v) Prevention of complications.

Q. 87. What are the processes by which healing takes place?

Ans. Following are the healing processes :

### (1) Healing by First Intention

This depends on the following conditions :

(a) Injury must be slight.

(b) Wounds should be clean, uninfected and cut wound.

*Process.* The fibrin formation from the blood takes place which brings and connects the separated ends. The gap is filled by fibroblast and endothelial cells of the connective tissue. Fibroblast absorbs the fibrin. Formation of new blood vessels takes place and restore the blood supply.

This type of healing occurs in those wounds which are made by operation. W.B.C. helps in the removal of bacteria, if any present.

### (2) Healing Under Scab

This depends on the following conditions :

(a) Wound should be wide.

(b) The ends are separate and they cannot come in contact to each other.

(c) Absence of infection.

*Process.* In presence of above factors the gap of the wound is filled with the formation of maximum amount of fibrin. The wound is covered by the fibroblast and endothelial cells by the formation of a white colour dense fibrous layer. The colour may be brown or black according to the skin colour. The formation of scar leads to formation of keloid.

### (3) Healing by Second Intention

Conditions :

(a) Wound is infected by pyogenic bacteria.

(b) Divided ends are separated forming a gap.

(c) Maximum amount of tissues are damaged.

*Process.* Healing occurs with the formation of granular tissue from the bottom of the wound. W.B.C., endothelial cells, fibroblast and new blood vessels are contained by granulation tissue. By phagocytic action W.B.Cs. remove the bacteria. Gap is covered by granulation. Blood supply to the part is restored by granulation tissue and endothelial cells start growing on the surface and formation of scar takes place as healing under scab.

### (4) Healing by Organisation

Condition :

Inflammatory case of serous membrane.

Such as—Peritonitis, pericarditis, pleurisy, periostitis, etc.

*Process.* A fibrinous layers form on the opposite surface firstly. Then healing occurs. Adhesion occurs in two surfaces which is followed by absorption of the inflammatory exudate so that adhesion is resolved and both layers become free again (in case of pleura, otherwise not). The endothelial cells, W.B.Cs. fibroblast and blood vessels are replaced by the granulation tissues. W.B.C. engulf the pathogenic micro-organism.

## (5) Healing by Resolution

Condition. It occurs in lobar pneumonea.

*Process.* Formation of exudate tpkes place which leads to destroy the toxic affect of pathogenic micro-organism. Polymorph of W.B.C. engulf the bacteria. Formation of pus occurs which is expectorated or absorbed healing takes place without formation of scar or granulation tissue.

## (6) Healing by Callus Formation

Condition : Fracture of bone.

*Process.* When rest is given to the part proliferation of osteoblast takes place leading to callus formation. Osteoblast acts like cementing and gives strength to the fractured bone.

## CHAPTER 3

# HELMINTHOLOGY

Q. 86. What do you mean by Helminthology ?

Ans. Helminthology is a branch of pathology which deals with the study of worms that produce diseases in men.

Q. 87. Give the character of Helminths.

Ans. Following are the character of helminths :

(1) Male worms are smaller than female.

(2) They enter in the human body in the form of ova or larvae.

(3) Helminths produce ova which can live outside the human body.

(4) Their life-cycle is of two types :
   (a) Intermediary host
   (b) In man.

(5) Each and every type of helminth has its own specific life-cycle.

Q. 88. What are the kinds of Helminths ?

Ans. All helminths are classified into two classes :

- (a) Nemat-helminths or non-segmented worm.
- (b) Platy-helminths or segmented flat worms.

Q. 89. What are the types of Nemat-helminths ?

Ans. They are of following types :

- (a) Oxyuris vernicularis, thread worms or pin worms.
- (b) Ascaris lubricoides or round worms.
- (c) Ankylostoma duodenale or hook worms.
- (d) Dracunculus medineusis guinea worms.
- (e) Trichuris trituria or whip worms.
- (f) Filaria bancrofti.
- (g) Trinhinella spiralis.

Other varieties are not so important.

Q. 90. What are the types of Platy Helminths ?

Ans. There are three types :

(1) *Costoda* :

- (a) Taenia Solium.
- (b) Taenia Saginata.
- (c) Hymenotepis Nana (H. Nana)
- (d) Echinococcus Granulosus.

(2) *Trematoda* :

- (a) Fasciola Hepatica.
- (b) Fasciola Buski.
- (c) Schistosoma Haematobium.

(3) *Turbellaria* :

They are not pathogenic.

## CHAPTER 4

# PROTOZOOLOGY

Q. 91. What are the main protozoal diseases of India ?

Ans. Following are the main protozoal diseases generally found in India :

(a) Malarial Fever.

(b) Amoebic Dysentery.

(c) Giardiasis.

(d) Kala-Azar

Q. 92. Give the name of different types of Malaria parasites ?

Ans. There are four types which mostly affect the man :

(a) Plasmodium Vivax.

(b) Plasmodium Falciparum.

(c) Plasmodium Ovale.

(d) Plasmodium Malariae.

Q. 93. Give the name of fever brought by different types of malaria parasites ?

**Ans.**

| Name of Malarial Fever | Name of Malarial Parasite |
|---|---|
| (1) Benign Tertian | Plasmodium Vivax |
| (2) Malignant Tertian | Plasmodium Falciparum |
| (3) Tertian Malaria | Plasmodium Ovale |
| (4) Quartan Malaria | Plasmodium Malariae. |

**Q. 94.** Give the name of causative organism of Amoebic Dysentery ?

**Ans.** Entamoeba Histolytica.

**Q. 95.** What are the stages of morphology of Entamoeba Hystolytica ?

**Ans.** There are three stages :

(a) Tropozoite form.

(b) Precystic form.

(c) Cystic form.

The metacystic form ends after the infection due to excystation.

**Q. 96.** In what form it is found in chronic case ?

**Ans.** The cystic form is found in the chronic dysenteric patient and in the carriers.

**Q. 97.** What are the mode of infection of entomoeba histolytica ?

**Ans.** The mode of spread are following :

(a) *Food.* It is spread by contaminated food.

(b) *Fly.* Flies are the most commonest source of contamination or infection.

(c) *Fomite.* The faecal matter may contaminate the food causing infection.

(d) *Drink.* The drinking water may be contaminated by the organism if it occurs in localised form. Then it causes local infection. If it affects the water supply from which mass population is connected then it causes epidemic of the disease. If the parasite is present to and fro in scattered form it causes endemic of the disease.

Q. 98. What is the difference between amoebic and bacillary dysentery ?

Ans. Difference between amoebic and bacillary dysentery may be divided into two parts :

(a) Macroscopic (Naked eye appearance).

(b) Microscopic appearance.

### Macroscopic

| *Amoebic Dysentery* | *Bacillary Dysentery* |
|---|---|
| (1) *Foecal Matter.* The foecal matters are always present together with mucous and blood. | (1) Foecal matters are hardly present. But mucous and blood are found. |
| (2) *Colour.* Dark red. | (2) Stool is bright red in colour. |
| (3) Not frank blood. | (3) Frank blood present. |
| (4) *Odour.* Offensive. | (4) Odourless. |
| (5) *Consistancy.* Liquid or formed. | (5) Viscid. |

(6) *Character.* No adherent to the container. (6) Adherent to the container.

(7) *Number.* Less. (7) More in number.

(8) *Quantity.* Less. (8) More in quantity

### Microscopic

*Amoebic Dysentry*      *Bacillary Dysentery*

(1) *Exudate.* Pus cells present in agglulinated form or clumps. (1) R.B.C. present in discrete manner.

(2) Eosinophile present. (2) Eosinophile absent.

(3) Macrophages (Inflammatory cells) absent. (3) Inflammatory cells present with injested R.B.C.

(4) Charcot layden crystal present. (4) Charcot layden crystal absent.

Q. 99. Give the name of Kala-Azar micro-organism ?

Ans. Leishmania Donovani.

# APPENDIX

The following figures indicate the relation of Fahrenheit to Centigrade degrees as required in clinical practice.

| Fahrenheit | Centigrade |
|---|---|
| 97·7°F. | 36·5°C. |
| 98·6 | 37·0 |
| 99·5 | 37·5 |
| 100·4 | 38·0 |
| 101·3 | 38·5 |
| 102·2 | 39·0 |
| 103·1 | 39·5 |
| 104·0 | 40·5 |
| 104·9 | 40·5 |
| 105·8 | 41·0 |
| 107·6 | 42·0 |
| 109·4 | 43·0 |
| 111·2 | 44·0 |
| 113·0 | 45·0 |
| 122·0 | 50·0 |
| 131·0 | 55·0 |
| 140·0 | 60·0 |

| Fahrenheit | Centigrade |
|---|---|
| 149·0°F. | 65·0°C |
| 158·0 | 70·0 |
| 176·0 | 80·00 |
| 185·0 | 85·0 |
| 194·0 | 90·0 |
| 203·0 | 95·0 |
| 212·0 | 100·0 |

**Conversion**

To convert c.m. $H_2O$ to mm of Hg, Multiply by 0·735.

To convert grammes into ounces, multiply by 10 and divide by 283.

To convert kilos into pounds, multiply by 1000 and divide by 454.

To convert Fahrenheit into Centigrade, substract 32, multiply the remainder by 5, and divide the result by 9.

To convert centigrade into Fahrenheit, multiply by 9, divide by 5 and add 32.

| To Change | To | Multiply by |
|---|---|---|
| Table-spoons | Cups | 0·0625 |
| ,, ,, | Cubic centimetres | 15·0 |
| ,, ,, | Dessert-spoons | 1·5 |
| ,, ,, | Tea-spoons | 3·0 |
| ,, ,, | Drops | 180·0 |
| Tea-spoons | Cubic centimetres | 5·0 |
| Tea-spoons | Cups | 0·02 |
| Tea-spoons | Drops | 60·0 |
| Grams | Ounces | 0·035274 |
| Gallons | Litres | 3·785 |

| | | |
|---|---|---|
| Feet | Centimetres | 30·48 |
| Cubic centimetres | Cubic inches | 0·061 |
| Cubic centimetres | Millilitres | 1·00 |
| Square centimetres | Square inches | 0·155 |
| Centimetres | Inches | 0·3937 |
| Ounces | Grams | 28·25 |
| Feet | Metres | 0·3048 |
| Inches | Millimetres | 25·4 |
| Millimetres | Inches | 0·03937 |
| Metres | Inches | 39·37 |

### Ideal Weights for Male ages 25 and over.

| Height (with shoes) Feet Inches | | Weight in pounds (as ordinarily dressed) | | |
|---|---|---|---|---|
| | | Small Frame | Medium Frame | Large Frame |
| 5 | 2 | 116—125 | 121—133 | 133—141 |
| 5 | 3 | 119—128 | 127—136 | 133—144 |
| 5 | 4 | 122—136 | 130—140 | 137—149 |
| 5 | 5 | 126—136 | 134—144 | 141—153 |
| 5 | 6 | 129—139 | 137—147 | 145—157 |
| 5 | 7 | 133—143 | 141—151 | 149—162 |
| 5 | 8 | 135—147 | 145—156 | 153—166 |
| 5 | 9 | 140—151 | 149—160 | 157—170 |
| 5 | 10 | 144—155 | 153—164 | 161—175 |
| 5 | 11 | 148—159 | 157—168 | 165—180 |
| 6 | 0 | 152—164 | 161—173 | 169—185 |
| 6 | 1 | 157—169 | 166—178 | 174—190 |
| 6 | 2 | 163—175 | 171—184 | 179—196 |
| 6 | 3 | 168—180 | 176—189 | 184—202 |

( 136 )

## Ideal Weight for Women ages 25 and over

| Height with shoes Feet Inches | | Weight in pounds as ordinarily dressed | | |
|---|---|---|---|---|
| | | Small Frame | Medium Frame | Large Frame |
| 4 | 11 | 104—111 | 110—118 | 117—127 |
| 5 | 0 | 105—113 | 112—120 | 119—129 |
| 5 | 1 | 107—115 | 114—122 | 121—131 |
| 5 | 2 | 110—118 | 117—125 | 124—135 |
| 5 | 3 | 113—121 | 120—128 | 127—137 |
| 5 | 4 | 116—125 | 124—132 | 131—142 |
| 5 | 5 | 119—128 | 127—135 | 133—145 |
| 5 | 6 | 123—132 | 130—140 | 138—150 |
| 5 | 7 | 126—136 | 134—144 | 142—154 |
| 5 | 8 | 129—139 | 137—147 | 145—158 |
| 5 | 9 | 133—143 | 141—151 | 149—158 |
| 5 | 10 | 136—147 | 145—155 | 152—166 |
| 5 | 11 | 139—150 | 148—158 | 155—169 |

(Metropolitan Life Insurance Company, Statistical Bureau, 1943).